Territory, Specialization and Globalization in European Manufacturing

Although traditional manufacturing (textiles, clothing, footwear, furniture, etc.) has been in decline in developed countries, it still represents an important part of European employment due to its labour-intensive character. Moreover, its geographical concentration particularly exposes certain regions of Europe to job loss as the industry declines. This book provides an explanation for the differences observed in the impact of globalization which is based on the influence of the territory and of the production specialization of the firms. The conclusions presented in the book are drawn from a detailed study of the Spanish textile-clothing sector.

The book highlights the intensity of the relationship between the organizational model of the territory where the firms are located (a high concentration of interrelated firms in a well-defined geographical area called an 'industrial district'), the specialization strategy implemented and the globalization of the economy. It also suggests the need to consider those factors as interdependent determinants of firm performance, particularly given the current trend for firms to simultaneously concentrate geographically and multilocalize domestically and internationally. The proposed methodology of analysis can be used to study other manufacturing sectors in other European countries.

A concise yet detailed volume, this is an invaluable contribution to the literature of economic and industrial globalization. It will prove invaluable to both researchers and policy-makers in the field.

Francisco Puig is Associate Professor in the Department of Management at the University of Valencia. **Helena Marques** is Associate Professor in the Department of Applied Economics at the University of the Balaeric Islands.

Routledge studies in global competition
Edited by John Cantwell
University of Reading, UK,
and
David Mowery
University of California, Berkeley, USA.

1 **Japanese Firms in Europe**
Edited by Frédérique Sachwald

2 **Technological Innovation, Multinational Corporations and New International Competitiveness**
The case of intermediate countries
Edited by José Molero

3 **Global Competition and the Labour Market**
Nigel Driffield

4 **The Source of Capital Goods Innovation**
The role of user firms in Japan and Korea
Kong-Rae Lee

5 **Climates of Global Competition**
Maria Bengtsson

6 **Multinational Enterprises and Technological Spillovers**
Tommaso Perez

7 **Governance of International Strategic Alliances**
Technology and transaction costs
Joanne E. Oxley

8 **Strategy in Emerging Markets**
Telecommunications establishments in Europe
Anders Pehrsson

9 **Going Multinational**
The Korean experience of direct investment
Edited by Frédérique Sachwald

10 **Multinational Firms and Impacts on Employment, Trade and Technology**
New perspectives for a new century
Edited by Robert E. Lipsey and Jean-Louis Mucchielli

11 **Multinational Firms**
The global–local dilemma
Edited by John H. Dunning and Jean-Louis Mucchielli

12 **MIT and the Rise of Entrepreneurial Science**
Henry Etzkowitz

13 **Technological Resources and the Logic of Corporate Diversification**
Brian Silverman

14 The Economics of Innovation, New Technologies and Structural Change
Cristiano Antonelli

15 European Union Direct Investment in China
Characteristics, challenges and perspectives
Daniel Van Den Bulcke, Haiyan Zhang and Maria do Céu Esteves

16 Biotechnology in Comparative Perspective
Edited by Gerhard Fuchs

17 Technological Change and Economic Performance
Albert L. Link and Donald S. Siegel

18 Multinational Corporations and European Regional Systems of Innovation
John Cantwell and Simona Iammarino

19 Knowledge and Innovation in Regional Industry
An entrepreneurial coalition
Roel Rutten

20 Local Industrial Clusters
Existence, emergence and evolution
Thomas Brenner

21 The Emerging Industrial Structure of the Wider Europe
Edited by Francis McGowen, Slavo Radosevic and Nick Von Tunzelmann

22 Entrepreneurship
A new perspective
Thomas Grebel

23 Evaluating Public Research Institutions
The U.S. Advanced Technology Program's Intramural Research Initiative
Albert N. Link and John T. Scott

24 Location and Competition
Edited by Steven Brakman and Harry Garretsen

25 Entrepreneurship and Dynamics in the Knowledge Economy
Edited by Charlie Karlsson, Börje Johansson and Roger R. Stough

26 Evolution and Design of Institutions
Edited by Christian Schubert and Georg von Wangenheim

27 The Changing Economic Geography of Globalization
Reinventing space
Edited by Giovanna Vertova

28 Economics of the Firm
Analysis, evolution and history
Edited by Michael Dietrich

29 Innovation, Technology and Hypercompetition
Hans Gottinger

30 Mergers and Acquisitions in Asia
A global perspective
Roger Y. W. Tang and Ali M. Metwalli

31 **Competitiveness of New Industries**
Institutional framework and learning in information technology in Japan, the U.S and Germany
Edited Cornelia Storz and Andreas Moerke

32 **Entry and Post-Entry Performance of Newborn Firms**
Marco Vivarelli

33 **Changes in Regional Firm Founding Activities**
A theoretical explanation and empirical evidence
Dirk Fornahl

34 **Risk Appraisal and Venture Capital in High Technology New Ventures**
Gavin C. Reid and Julia A. Smith

35 **Competing for Knowledge**
Creating, connecting and growing
Robert Huggins and Hiro Izushi

36 **Corporate Governance, Finance and the Technological Advantage of Nations**
Andrew Tylecote and Francesca Visintin

37 **Dynamic Capabilities Between Firm Organisation and Local Systems of Production**
Edited by Riccardo Leoncini and Sandro Montresor

38 **Localised Technological Change**
Towards the economics of complexity
Cristiano Antonelli

39 **Knowledge Economies**
Innovation, organization and location
Wilfred Dolfsma

40 **Governance and Innovation**
Maria Brouwer

41 **Public Policy for Regional Development**
Edited by Jorge Martinez-Vazquez and François Vaillancourt

42 **Evolutionary Economic Geography**
Location of production and the European Union
Miroslav Jovanovic

43 **Broadband Economics**
Lessons from Japan
Takanori Ida

44 **Targeting Regional Economic Development**
Edited by Stephan J. Goetz, Steven C. Deller and Thomas R. Harris

45 **Innovation, Knowledge and Power in Organizations**
Theodora Asimakou

46 **Creativity, Innovation and the Cultural Economy**
Edited by Andy C. Pratt and Paul Jeffcutt

47 **Co-opetition Strategy**
Giovanni Battista Dagnino and Elena Rocco

48 **Knowledge Intensive Entrepreneurship and Innovation Systems**
Evidence from Europe
Edited by Franco Malerba

49 **Innovation in Complex Social Systems**
Edited by Petra Ahrweiler

50 **Internationalization, Technological Change and the Theory of the Firm**
Edited by Nicola De Liso and Riccardo Leoncini

51 **Territory, Specialization and Globalization in European Manufacturing**
Francisco Puig and Helena Marques

Territory, Specialization and Globalization in European Manufacturing

Francisco Puig and Helena Marques

Routledge
Taylor & Francis Group
LONDON AND NEW YORK

First published 2011
by Routledge
2 Park Square, Milton Park, Abingdon, Oxon OX14 4RN

Simultaneously published in the USA and Canada
by Routledge
711 Third Avenue, New York, NY 10017, USA

Routledge is an imprint of the Taylor & Francis Group, an informa business

Typeset in Times New Roman by Wearset Ltd, Boldon, Tyne and Wear

British Library Cataloguing in Publication Data
A catalogue record for this book is available from the British Library

Library of Congress Cataloging in Publication Data
A catalog record for this book has been requested

ISBN13: 978-0-415-55206-6 (hbk)
ISBN13: 978-0-203-84480-9 (ebk)

Contents

Figures

Tables

Preface

The idea of a book on the interrelation between territory, specialization and globalization that would bring to light the challenges faced by European traditional manufacturing came up back in 2008, when Helena was Senior Lecturer at the University of Manchester and Francisco was taking a research visit at the university. He shared with Helena an interest in the textile-clothing industry, which still has a strong presence in their native countries, Spain and Portugal. In 2005, when the World Trade Organization removed the 50-year-old quota system from the worldwide textile-clothing trade, Francisco had conducted a survey of the textile-clothing industry looking at the moderating effect of the territory and specialization on firm performance, which was the beginning of interesting and fruitful research. At the same time, Helena had published a number of articles in scientific journals on themes of international trade, economic integration and economic geography using European evidence. The present book tries to place the lessons learned from their respective research backgrounds in the context of traditional manufacturing in Europe in an era of globalization.

The authors wish to thank Pervez Ghauri, who invited Francisco to visit the University of Manchester and so made the present book possible. Francisco also thanks his family for their support ever since the groundwork that kickstarted this book was initiated and the textile-clothing association ATEVAL for their collaboration, which made the survey carried out in Spain possible. Thanks are also due to colleagues at the University of Valencia, particularly José Pla-Barber, whose comments and suggestions on initial versions of the work allowed it to take the form it now has. Helena is indebted to her family for their invaluable support since this project was started, as well as to the University of Manchester (2008–09) and the University of the Balearic Islands (2009–10) for allowing her the time to work towards this book. She hopes that the book will be of interest to the traditional manufacturing industry in other countries, particularly in Portugal, which, in the EU context, remains strongly specialized in textile-clothing and other traditional manufacturing.

Francisco Puig and Helena Marques
Valencia and Palma de Mallorca

1 Introduction

1.1 Background and motivation

The present book studies the recent evolution of the interrelations between territory, specialization and globalization that have been observed in European traditional manufacturing in the current, very competitive environment. The topic is broad in its applications and so the book focuses on one particular geographical space (Europe) and on one particular industrial sector (traditional manufacturing). Within this context, an empirical analysis of the textile-clothing industry in Spain is presented in detail. The reasons why this case is representative of the current situation and of the challenges facing European traditional manufacturing will be briefly explained in this overview. Among them is the fact that the textile-clothing industry in Southern Europe (Italy, Spain and Portugal being prime examples) has been organized into networks of very specialized small and medium enterprises (SMEs) embedded in the territory in which they are located. This traditional model of organization is, however, being challenged by the globalization process.

In the mid-1990s, the rumours of the end of the World Trade Organization's Agreement on Textiles and Clothing led to fears of the dismantling of European traditional manufacturing, comprising to a large extent textile-clothing SMEs. The territorial embeddedness of the industry caused concern for the economic and social future of the European regions that were still heavily specialized in traditional manufacturing. Southern European countries such as Italy, Spain and Portugal would be among the most adversely affected. Authors such as Giacomo Becattini (1979, 2004) and Michael Porter (1990, 1998) studied successful experiences in various traditional manufacturing sectors within the territorial network framework. The social and organizational characteristics of the textile-clothing sector and the competitive advantages brought by the network model are particularly visible in the various industrial districts that have been identified throughout Europe, especially in Italy and Spain.

Some of the previous literature on industrial districts has found a certain degree of firm heterogeneity in the participation and exploitation of territorial advantages, sometimes with contradictory conclusions (Camisón, 2004; Molina and Martínez, 2004; Giuliani, 2007). The heterogeneity of firm-level responses to globalization was also a feature of the results of the survey distributed to textile-clothing firms in 2005, which are examined in Chapters 5 and 6. A deeper study of the underlying causes of that heterogeneity was required. Some of the differences highlighted by previous literature were partly dependent on the use of a range of methodologies and samples, without taking into account the competitive environment of the firms, that is, the territory where they were located. However, this book shows that another reason for heterogeneity is the nature of firm specialization within the same traditional manufacturing industry. More concretely, SMEs that specialize in the main activity of an industrial district tend to be more optimistic about their environmental conditions and their capabilities – although with more heterogeneous evaluations of the explanatory factors – than the rest of the firms in that industrial district. These observed differences force a reevaluation of strategic prescriptions to firms.

In this context, the book pays particular attention to the definition of geographical areas and industries. By choosing a certain industrial district and well-defined subsectors as units of analysis, the book is able to study the joint influence of the territory and of the specialization pattern on heterogeneous traditional manufacturing firms. Two aspects of the firms are studied: managerial perceptions and internationalization strategy. Managerial perceptions are important because they condition strategic decisions (Chen *et al.*, 2007). There is evidence of this link in the literature and more directly through the research previously carried out in Spain with managers, workers and industry experts, which is reported in Chapters 5 and 6. Moreover, managerial perceptions change with the territory and specialization patterns. On the other hand, economics and strategy research has recognized that the corporate strategy of internationalization is a key response to globalization (Pla and León, 2004; Daniels *et al.*, 2007). The exploitation of new markets offers important advantages to firms, through growth, risk diversification, profitability, scale economies, tax advantages, lower labour costs and lower competition levels. At the same time, the increasingly global orientation of traditional manufacturing is changing the organizational model of industrial districts and fostering doubts about its future (DeMartino *et al.*, 2006; Zucchella, 2006).

In conclusion, this book shows that firms within the same traditional manufacturing industry (in this case, textile-clothing) are heterogeneous, especially the SMEs. The firms sampled exhibit differences in managerial perceptions and internationalization strategies that are related to their location in an industrial district (or outside) and to their specialization patterns (as subsectors of an activity). Consequently, asymmetries in the impact of globalization are expected among firms and such asymmetries need to be recognized and taken into account by managers and policy-makers.

1.2 Current challenges

The current levels of globalization, competition and deregulation have created a changeable, dynamic and uncertain environment that has been forcing changes in the organization of production (Sammarra and Belusi, 2006; Puig *et al.*, 2009). Firms of various dimensions and nationalities have had to change their strategies in order to adapt to the environment, with important economic and non-economic costs. Many times, the alternative to changing may simply be to exit the activity. Two possible strategies are the specialization in those tasks that are less susceptible to globalization (typically those that require higher skill levels and are more intensive in R&D) and the adoption of network models such as the 'industrial district'. In this context, an industrial district is defined as a geographical agglomeration of SMEs operating within a single industry that have peculiar organizational and productive characteristics, are socially embedded and have a global orientation (Becattini, 1979).

There is a strong relationship between industrial districts, specialization, internationalization and globalization that suggests the need to consider the behaviour of firms under all those aspects. The location in an industrial district and an adequate specialization pattern can favour innovation processes and openness to foreign markets, at the same time increasing productivity and conditioning the strategic decisions of the managers. The study of industrial districts can contribute to an explanation of the different levels of competitiveness shown by the same subsector in different locations, which in turn is influenced by the presence of exceptionally competitive firms. In this way, the industrial district arises as an organizational system that allows the understanding of how traditional manufacturing responds to current challenges. As pointed out by Pyke and Sengenberger (1992), the greatest problem of manufacturing SMEs is not their size, but being alone. On the other hand, the internationalization strategy has been emerging as a key response to globalization. In Europe, internationalization has been supported at the governmental level, with the implementation of several public support programmes for exporting, innovation and worker training (Oxelheim and Ghauri, 2004). At the private level, industry associations have promoted the understanding of the new scenarios and the consequent adaptation of firms. Research centres and universities have also collaborated in R&D programmes to enhance competitiveness.

Still, the viability of industrial districts and traditional manufacturing in Europe has been called into question, and it remains to be defined what role, if any, governments can have in supporting them (Zourek, 2007). These issues will be examined in the context of the European textile-clothing industry by establishing the influence of the territory and specialization on managerial perceptions and internationalization strategies of SMEs, viewed as heterogeneous firms.

1.3 Overview of the book

The book is made up of seven chapters. The next three establish the background on networks and industrial districts (Chapter 2), internationalization (Chapter 3) and the textile-clothing industry (Chapter 4). They provide the context for studying textile-clothing as a traditional manufacturing industry and the working hypotheses to be transposed to the territory examined in the following two chapters (Chapters 5 and 6). The final chapter (Chapter 7) brings both parts together with final conclusions, limitations and implications for future research. A brief overview of Chapters 2–6 is provided below.

Chapter 2 reviews network models of industrial organization and their role in defining a territory. It then focuses on a particular type of network model, the industrial district, and analyses the organizational and strategic implications of district location.

Chapter 3 reviews the literature on internationalization and the impact of industrial location on internationalization patterns. The chapter argues that globalization promotes the delocalization of production tasks and the district location inhibits them. As a consequence, the internationalization strategy that has commonly been carried out as a response to globalization, based on imports and delocations, tends to be weakened for district firms. Nevertheless, the influence of the industrial district is visible in other aspects of international activity, such as exporting propensity and intensity.

The tension between the local and the global contexts is particularly acute in traditional manufacturing. Chapter 4 characterizes the current situation of the textile-clothing industry as a representative case, first in Europe, then in Spain, finally focusing on the Comunidad Valenciana (NUTSII region ES52). Three important conclusions emerge from this chapter: (1) traditional manufacturing faces a structural crisis; (2) the textile-clothing industry in the Comunidad Valenciana has specialized in home textiles, which are labour-intensive and very low in R&D; and (3) it is possible to identify a strong agglomeration of textile-clothing SMEs in the area of Alcoi-Ontinyent, which will be coined as the Valencia textile-clothing industrial district.

Having reviewed the theoretical background, characterized traditional manufacturing in Europe and summarized the main implications derived from Chapters 2–4, Chapters 5 and 6 are dedicated to the analysis of a questionnaire distributed to and returned by approximately 200 textile-clothing firms affiliated to the main textile-clothing association in Spain (ATEVAL), most of which are located in the Comunidad Valenciana, around the industrial district of Alcoi-Ontinyent.

Chapter 5 proposes a new methodology of identification of the geographical area of an industrial district at the municipality level. It then describes the population of firms to which the questionnaire was distributed, the content of the questionnaire and its distribution. Finally, some descriptive statistics are presented for the sample, as well as some preliminary relationships between the variables defining the sample characteristics (size, age, location, subsector, international activities). There is a strong relationship between the home-textile subsector and the district location, but a weak relationship between them and international activities. Further, district location and subsector specialization are highly independent of firm size and age.

Chapter 6 analyses the strategic environment of sample firms in order to understand the managerial perceptions of the situation and evolution of the textile-clothing industry. It then identifies some key aspects of their business activity: threats, opportunities, weaknesses, strengths, strategic options and public policy demands. The second part of the chapter provides a statistical analysis, which includes means testing, analysis of variance and the use of ordinal logit models to test a number of hypotheses regarding the influence of district location and subsector of specialization on managerial perceptions and international activity.

The present book views the textile-clothing industry in the Comunidad Valenciana as representative of traditional manufacturing in Europe. The authors would like to encourage other researchers to extend the analysis to other regions and countries, as well as to other traditional manufacturing industries. The consolidation of the conclusions presented in this book would be beneficial for managers and policy-makers alike.

2 Networks and industrial districts

Business activity is immersed in a complex and varied conglomerate of relationships or networks that firms need to maintain and develop so they can adapt effectively to the constant changes they face in the competitive environment (Jones *et al.*, 1997). This observation has been noted by several fields of research – economics, geography, sociology, business studies and political science, among others – and materialized in a growing interest in firms' networks.

There are two main strands of research on this topic (Möller and Svahn, 2003): one sees networks as the environment in which economic activity takes place ('network of organizations'); another, centred on the characteristics and effects of networks, and the density, multiplicity and reciprocity of relationships among its members, deals with 'network organization'. Within the second strand it is possible to identify three main perspectives according to the network elements, their relationships and the importance placed on the territory: organizational (Ghosal and Bartlett, 1990), social (Granovetter, 1985) and strategic (Gulati *et al.*, 2000).

The strategic perspective places a special emphasis on territorial networks, considering them as a way of organizing business activities that can notably increase firm competitiveness. One of the most studied representations of territorial networks is the geographical concentration of firms within the same industry. It is believed that such a model generates important benefits for participating firms, embodied in the existence of external economies (Krugman, 1991a; Parr, 2002a, 2002b). From a conceptual point of view, when that concentration results in the grouping of a sufficiently high number of firms, particularly SMEs, we talk about clusters and/or industrial districts.[1]

It has been argued that the ultimate outcome of belonging to a cluster or industrial district is higher competitiveness. The influence of the territory on the firms located therein has been tested for structural characteristics of the firms, such as economic performance (Porter, 1990; Bagella, 2000; Puig *et al.*, 2009). These studies have allowed the identification of some of the effects of localization. However, there are some limitations in this line of study with respect to methodology and scope (Paniccia, 1998; Becattini, 2004). In particular, the analysis of the influence of the territory on firms' characteristics or managerial perceptions is still scarce.

In this context, it is important to start by understanding both the background and conceptualization of industrial districts as a type of territorial network, and some of the implications of district location for business activity. The first of the following four sections analyses the main types of configurations of networks of firms. The second revises and discusses the industrial district as a unit of analysis in economic and managerial terms. The third studies the main strategic implications of that organizational model and establishes the methodological basis for analysis. The fourth section summarizes the main conclusions of the chapter.

2.1 Inter-organizational relationships and networks: a configuration typology

Inter-organizational relationships and networks have been the subject of considerable attention. The motivation appears to have come from the benefits offered by networks, such as access to external resources, risk-sharing and risk-pooling, and access to external information. Nevertheless, membership of a network also brings a risk of strategic dependency. In any case, researchers have studied interactions using different levels of analysis and perspectives, leading to the development of a 'terminological jungle'.

The analysis can be carried out at the dyadic level, looking at the relationship between two organizations, and at the multiple level, which considers relationships between more than two organizations, including all the connections that make up the organizational system, seen as the set of coordinated actions of various interconnected organizations (García, 1996). The latter level of analysis has been referred to in the literature as *network*, although many alternative definitions have been proposed, some of which are summarized in Box 2.1.

Box 2.1 The concept of 'network'

- A network is a group of people connected by relationships that create attachments between its members (Aldrich and Zimmer, 1985).
- Networks are a form of organization between the hierarchy (integrated firm) and the market (purely external relationships) (Thorelli, 1986).
- Networks can be seen as a form of organization that can be used by managers and entrepreneurs to position their firms at higher competitive levels (Jarillo, 1988).
- A network can be seen as a set of nodes or positions occupied by independent firms, linked by a set of relationships that imply agreements between those firms (Menguzzato, 1992: 40).
- Networks are groups of firms coordinated by market mechanisms instead of hierarchical chains (Miles and Snow, 1992).
- Networks can be seen as a group of firms, among which there is no relationship of subordination, that maintain several cooperative links, through which they can perform jointly coordinated actions (García, 1996).

- A network can be defined as two or more organizations enveloped in a long-term relationship (Snodgrass, 1993).
- A network is a set of trust-based relationships between people that are part of the same social group (Casson, 1998).
- A network is the set of formal and informal relationships around the entrepreneur (Vázquez, 1999).

A first look at alternative definitions shows that the concept of *network* is far from clear. It is an abstract notion that makes reference to different types of nodes (people, firms or other institutions) and to the various relationships that link these nodes (Knoke and Kuklinski, 1983). However, looking more closely at those definitions, they can be organized around three main perspectives, according to the elements considered and the characteristics of the relationships between them (Grandori and Soda, 1995; Ebers and Jarillo, 1998; Oliver, 1998; Inkpen and Tsang, 2005). The first level is organizational or intrafirm, which includes distinct elements internal to the firm. The second level is social and implies the participation of entrepreneurs and their environment. The third level considered is strategic or interfirm and includes different types of organizations.

2.1.1 The organizational perspective

The organizational perspective originates in the theory of transaction costs (Williamson, 1979). According to this line of research, the main justification for the existence of networks lies in the efficiency gains they bring – pecuniary externalities (Krugman, 1991a) that allow savings through the proximity to upstream and downstream firms. A great deal of the research on this perspective has investigated the origins and elements of the networks rather than their impact. Perrow (1992) argues that the causes of the emergence of networks lay on the adoption of flexible production and the questioning of the efficiency model of the large integrated firm. On the other hand, Ghoshal and Bartlett (1990), among others, study firms that carry out international activities while remaining under the same decisional unit, such as the multinationals.

The elements that constitute the network differ according to the disaggregation level used for the value-chain[2] activities. The literature has established a distinction between the 'federal' and the 'tree' organizational models. The federal model has been applied by those firms that had to adjust their structures to new organizational demands due to the corporate strategies adopted (Fernández, 1994). The tree model, on the other hand, reflects operational strategies. In this case, the network is formed through the establishment of agreements, typically of subcontracting, between distinct and autonomous organizational units. The process is initiated when the main firm sources, totally or partially, an activity from another firm or firms (Linares and Pla, 2009). This sourcing process is translated into a complex network of relationships that allow greater flexibility and capacity of adaptation to rapidly changing environments. It is also common to cover low-skill activities with flexible labour on temporary contracts, leaving high-skill activities to specialized workers on permanent contracts.

In sum, under this perspective the networks of firms are a response to the challenges of a changing environment. Independent of the relationships between the elements forming the network, the main focus here is on the study of the differences between the organizational forms adopted. Hence, its main interest resides in the explanatory power offered to understand the responses of firms using their organizational design.

2.1.2 The social perspective

The study of networks from the social perspective sees the firm as an economic agent that is embedded in a complex web of relationships with other economic agents. This line of research focuses more on the relationships themselves than on the characteristics of each element (Gulati and Garciulo, 1999). Its main theoretical background is network theory, which suggests that opportunities are provided to economic agents by their position in a network of relationships and by the content of such relationships (Oliver, 1998).

One of the most influential concepts of network theory is that of embeddedness, according to which business activity is embedded in social and business relationships (Granovetter, 1985; Staber, 2007). The economic importance of social capital has long been recognized and there is presently an important base of research on the importance of social capital for both individuals and firms (in the context of firms, see Molina (2005), Annen (2001), Cooke *et al.* (2005), Cooke and Wills (1999), Woolcock and Narayan (2000), Ostrom (1996), La Porta *et al.* (1997), Akerlof and Kranton (2000), North (1991) and Sachwald (1998)). Szarka (1990) and Curran *et al.* (1993) have highlighted the importance of the social environment (akin to the concept of social capital) for the development of SMEs. Aldrich and Zimmer (1985) also mention the importance of networks for entrepreneurs.[3] Similarly, Hite and Hesterly (2001) state that the importance of networks varies along the lifecycle of the firm and is particularly important at its birth and infancy.

The nature and evolution of networks can also be explained through the theory of endogenous growth using the concept of external effects that spring from economies of scale and the pooling of human and social capital through geographical agglomeration (see Baldwin and Forslid (1997) for a simple core–periphery model that incorporates elements of both new economic geography and endogenous growth). In this context, the various types of relationships established between the agglomerated firms provide the mechanisms for the processes of growth and structural change of the local and regional economies (Baldwin *et al.*, 2003). Therefore, the network framework becomes a useful instrument to explain several phenomena such as the dynamics of firm behaviour and the local production systems.

The social approach identifies several types of networks: (1) with respect to the formal and informal relationships established between firms; (2) according to the nature of transactions, whether of information (technical relationships) or goods and services (commercial transactions); (3) according to the actors involved, whether they be (a) personal networks, which provide personal information and resources, or (b) firms' networks, which provide business information, technical assistance, financial resources and strategic alliances (Vázquez, 1999).

In turn, Bramanti and Senn (1993) classify networks according to the degree of formalization of relationships, or visibility, and according to the degree of commitment or complexity. The relationships characterized by low formalization and low commitment, such as personal networks and informal firms' networks, are implicit or tacit. On the contrary, explicit and codified networks are based on strong formalized commitments by those involved.

To summarize, the social perspective highlights the social context of the firms. The knowledge of the space in which economic activity and its relationships take place is a key aspect in explaining the functioning and the business dynamics of the territorial organizational models.

2.1.3 The strategic perspective

The strategic perspective on firms' networks is based on the strategic consequences of belonging to a network, especially for the smallest firms (Gulati *et al.*, 2000). The theoretical background is based on the assumption that networks result from an evolutionary process of the environment (Oliver, 1998; Dangayach and Deshmukh, 2001).

The bulk of research in this area follows one of the following lines: (1) the estimation of the impact of network membership on firm performance (Human and Provan, 1997); (2) the determination of the impact of network membership on strategic development (Johanson and Mattson, 1988; Kogut, 2000); and (3) the valuation of the firm's ability to acquire competitive capabilities (McEvily and Zaheer, 1999), such as the importance of networks as a factor conditioning internationalization (Welch and Welch, 1996).

In this context, the classification of networks is based on the degree of specialization of production activities and degree of autonomy of participating firms (Fernández, 1996), and on the governance and network characteristics such as intensity, scope and commitment (Langlois and Robertson, 1995). Figure 2.1 represents the second classification, with six possible types of network: holdings, multidivisional firms, *kaisha*, risk capital networks and territorial base networks. Within the latter, there are both Marshallian networks (clusters) and Becattinian networks (industrial districts). According to the classification represented in Figure 2.1, the common characteristic of both types of territorial network is the low integration (high interdependency) of its firms, while they differ in the level of coordination (intensity, scope or commitment) between the participants.

In sum, the strategic perspective takes into account simultaneously the power relationships, the degree of specialization and the territory of the participating firms. By doing this it allows the identification of the realities characterized by the presence of different types of firms, provides an explanation of non-hierarchical models and an understanding of their implications for the activity

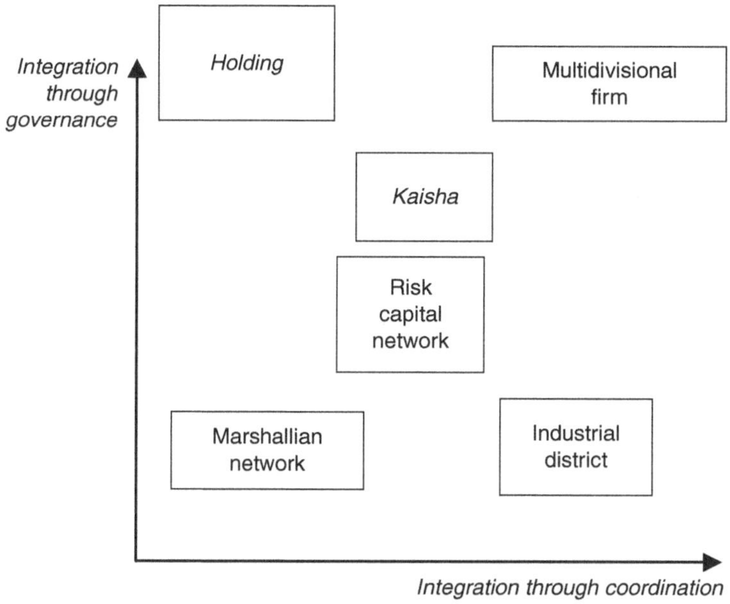

Figure 2.1 Strategic network (source: adapted from Langlois and Robertson (1995)).

and strategy of firms. As shown in Table 2.1, the strategic perspective is distinguishable from the other two in that the main participants in the network are firms that relate to one another through mutual strategic dependence.

The levels of dependency highlighted by the strategic perspective suggest various stages and types of territorial networks whose extensions are observable in any economic activity. However, within the manufacturing sector, the most relevant are the territorial networks with high autonomy of participating firms.

2.1.4 Typologies of territorial networks

The representation of territorial concentrations of manufacturing firms reflects the combinations and stages of evolution adopted as a response to technological and market changes (Sölvell, 2009). Consequently, there are many models of geographical concentration of industries. Amin and Robins (1992) present a typology around the participating firms, their activity and network purposes, which establishes a useful characterization. Brusco (1992) identifies four models of territorial agglomerations of SMEs: traditional artisan, dependent subcontractor, first-generation industrial district and class-II industrial district. The first two models were the result of the strategies of production decentralization of large firms. In both types of industrial districts firms show low dependency, although only class-II districts benefit from government support (Brusco, 1992).

Table 2.1 Perspectives of network research

	Network elements	*Characteristics of relationships*	*Object of research*	*Some models*
Organizational	Organizational units	Power-based Structured	Network forms	Federal Tree
Social	Firms and individuals	Trust-based Unstructured	Relationships: formalization, intensity and complexity	Personal Informal
Strategic	Independent firms	Cooperation Variable structure	Strategic consequences	Clusters Industrial districts

Markusen (1996) identified five types of agglomerations according to their orientation and governance structures: Marshallian industrial districts, Italian industrial districts, hub-and-spoke districts, satellite industrial platforms and state-anchored districts. The first two are based on local SMEs that benefit from skilled labour pooling and knowledge externalities, although the structures of coordination and cooperation are much more developed in the Italian model. The hub-and-spoke districts are those where the agglomeration is dominated by one or a few large firms, which share few territorial elements. The satellite industrial platform corresponds to a concentration of affiliates of a large firm, typically externalized to take advantage of lower costs. Finally, the state-anchored district is a large technological park built around one or more public-sector institutions.

On the other hand, Altenburg and Meyer-Stamer (1999) and McCormick (1999) have looked essentially at the efficacy of industrial policies. They highlighted that the various technological stages and the different elements that characterize the territory create different organizational models that require alternative economic policies. Altenburg and Meyer-Stamer (1999) distinguish among survival clusters, advanced clusters and technological clusters. The first type is composed of SMEs that, due to the absence of liaison and support services, require public intervention to rescue them. The other two types are dominated by large firms located outside the cluster, assuming that their creation through policies of R&D and skill training will foster local growth and become self-replicating. McCormick (1999) proposes three types of geographical concentrations of industry according to the processes of industrialization: (1) the pre-industrial cluster, defined as the one that initiates development; (2) the industrialized cluster, which facilitates specialization; and (3) the complex industrial cluster, which produces competitively for a variety of markets.

Finally, Brenner (2004) distinguishes five types of geographical concentrations: (1) Marshallian industrial districts; (2) Italian industrial districts; (3) industrial clusters; (4) technological districts, composed of affiliates, outsourced firms and decentralized units of a large firm, located in particular industrial areas; and (5) *milieu* or innovating environments, areas characterized by strong innovation that results from frequent contacts and collective actions by a set of institutions such as technological centres and universities (Baptista and Swann, 1998; Bell, 2005).

In summary, as Table 2.2 shows, a cluster or an industrial district corresponds to a specific type of territorial concentration and does not follow a unique model. Nevertheless, with the purpose of establishing the bases for its study as a unit of analysis, it is necessary to distinguish the concept of 'cluster' from that of 'industrial district' (Marshallian or Italian): all industrial districts are clusters, but not all clusters are industrial districts (Schmitz, 1995).

The concept of industrial, regional, economic or business clusters originates in the work of Marshall (1890) on industrial agglomeration. Both 'cluster' and 'industrial district' are terms used to identify territorial networks of firms and production systems (Belussi and Caldari, 2009). Indeed, industrial districts have been called models of industrial organization (Piore and Sabel, 1984), a way of organizing production (Storper, 1997) or a type of firms' network (Langlois and Robertson, 1995). Saxenian (1994) refers to Silicon Valley and Route 128 as examples of both industrial districts and clusters. The pottery concentration of Sassuolo (Italy) is described by Porter (1990, 1998) as a regional/industrial cluster, while most Italian authors call it an industrial district. Krugman (1991b) points out the case of the automobile industry in Detroit.

Table 2.2 Types of clusters

Authors	*Perspective*	*Proposed typology*
Brusco (1992)	Evolution of concentrations of SMEs over time	Traditional artisan Dependent subcontractor First-generation industrial district Class-II industrial district
Markusen (1996)	Participating firms and governance structures	Marshallian industrial districts Italian industrial districts Hub-and-spoke districts Satellite industrial platforms State-anchored districts
Altenburg and Meyer-Stamer (1999)	Technological stages of agglomeration	Survival clusters Advanced clusters Technological clusters
McCormick (1999)		Pre-industrial clusters Industrialized clusters Complex clusters
Brenner (2004)	Integrates the previous ones	Marshallian industrial districts Italian industrial districts Industrialized clusters Technological districts *Millieu* or innovation environments

Effectively, an industrial cluster is a geographical concentration of industrial firms. Similarly to an industrial district, that agglomeration is located in an area populated by firms operating in a single industry with its downstream and upstream components. It is thus characterized by specialization and relationships between firms. Following Gordon and McCann (2000) and Becattini (2002), the concepts of cluster and industrial district are used interchangeably when agglomerations are studied from the wider perspective of geographical organization of industry and determinants of competitiveness; when discussing the social and organizational characteristics of those agglomerations, only the concept of industrial district will be used.

2.2 Industrial districts as territorial units of analysis

2.2.1 Background

Marshall (1890) pointed out the advantages brought by industrial districts to the participating firms. He suggested that those advantages originated from external economies linked to the division of labour among specialized units and a geographical space with certain geographical, social and cultural characteristics (Belussi and Caldari, 2009). He further distinguished between economies internal to the firm and economies external to the firm but internal to the industry. Internal economies result from the economies of scale and scope obtained by increasing production capacity, reorganizing the production structure and increasing management efficiency. External economies depend on the general development of the industry to which the firm belongs and consist mainly of pecuniary externalities (Krugman, 1991a) due to the geographical concentration of downstream and upstream firms and to locating close to workers and markets. Naturally, such pecuniary externalities diminish with reductions in transport costs, although arguments linked to the pooling of skilled labour and knowledge spillovers still remain and become increasingly important (Marques, 2005).

Marshall (1890) had initially introduced the concept of external economies to explain two related aspects. First, external economies determine the location of an activity in a territory.[4] Together with physical and economic geography, demand conditions and history, external economies can explain the geographical concentration of firms. This is similar to the distinction between 'first nature' and 'second nature' advantages established by Krugman (1993). Second, external economies are a source of competitive advantage for firms. This second aspect is directly derived from the first, as rational firms would not choose to agglomerate unless they would perceive advantages in doing so.

The external economies obtained with agglomeration are the basis for gains from this production model. As Marshall pointed out, agglomeration creates an environment conducive to innovation and facilitates access to specialized machinery and skilled workers. Following the seminal work by Krugman (1991a), the geographical economics literature looked at agglomeration economies as pecuniary externalities, that is, the reduction in each firm's costs obtained from concentrating the industry in a single location. In doing this, each firm gains and so does the industry as a whole. The origin of the savings lies on with the lower cost of inputs obtained from local suppliers, due to lower transport costs and knowledge spillovers. In Krugman (1991a) and the ensuing literature, the price index of inputs is defined as a weighted average of the price of inputs sourced from all the possible locations, with inputs from other locations subject to transport costs. Under this formulation, the savings made possible by agglomeration become clear: a firm's unit cost decreases with the price index of inputs, and so with agglomeration as well.[5]

The geographical economics literature explains that economic activity shows a tendency to concentrate geographically and the regions with favourable initial conditions are those where industry will agglomerate. It has, however, been shown that not all industries show the same tendency to agglomerate. Steinle and Schiele (2002) distinguish a set of necessary conditions of a technical nature – divisibility of the production process and transportability of the final product – as well as a set of sufficient conditions: a long value chain, multiple competencies, innovation networks and a volatile market. The industries that meet these conditions will show a greater tendency to agglomerate. Due to its characteristics, the textile industry, in particular, is among those most prone to agglomeration (Venables, 1999; Baldwin *et al.*, 2000; Brulhart *et al.*, 2004; Head and Mayer, 2006; Marques and Metcalf, 2009).

The economics tradition discusses agglomeration economies as external economies, where the advantage for firms from being close to other firms comes not from their internal dimension, but from the dimension of the industry in the same geographical space. Nevertheless, Parr (2002a) sees agglomeration economies as being both external and internal to the firm, where the latter depend only on the characteristics of that particular firm, independently of the number of firms concentrated in the same geographical space. He further considers that both internal and external economies can be economies of scale (dependent on the volume of output), scope (dependent on the number of product lines) or complexity (dependent on the level of vertical integration). This typology is summarized in Table 2.3.

Localization economies originate from the common location and interdependence of independent firms with specialization in the same industry (e.g. clusters and industrial districts). Urbanization economies also result from the common location of firms, in this case belonging to different industries, which benefit

Table 2.3 Typology of agglomeration economies

Type	*Internal economies*	*External economies*
Scale	Horizontal integration	Localization
Scope	Lateral integration	Urbanization
Complexity	Vertical integration	Complex activity

Source: Parr (2002a).

from the high level of economic activity in a certain location (e.g. industrial areas). Complex activity economies are related to a location occupied by firms that belong to the same value chain and benefit from locating close to upstream and downstream firms (e.g. suppliers of automobile parts that agglomerate close to their downstream client).

Although agglomeration can bring important benefits to participating firms, such as the external economies found in industrial districts, the impact is not always positive and linear.[6] The geographical agglomeration of industry creates demand for immobile, or less mobile, production factors, such as labour and land, driving up prices and, if labour and land become sufficiently expensive in time, eroding the advantages of agglomerating. Additionally, there are other congestion costs that have been well documented by economic geographers, such as traffic congestion and the environmental costs of pollution (Paniccia, 1999). A final issue that has been widely debated, particularly in the context of European integration, is that of asymmetric shocks: the impact of a negative asymmetric shock affecting only a particular industry is higher in more specialized regions. A recent example is that of the liberalization of the world textile trade by the WTO in 2005, which particularly affected the EU regions that were more specialized in textile-clothing production.

On the other hand, it has also been argued that agglomeration in districts can protect SMEs in times of crisis by giving very specialized firms an opportunity to combine certain common functions. In particular, specialization in a particular process or product may allow a firm to diversify into new product lines (Schmitz, 1995).

2.2.2 Evolution and terminology

Since the 1970s there has been a consistent and growing research effort on industrial districts, which is related to the economic success of certain industrial regions of advanced economies that experimented with the district model, in particular the centre and north of Italy (Dunford, 2006). At the same time, the district phenomenon has been studied in developing countries for various industries.[7] According to Bellandi (2002), the success of industrial districts was due to environmental factors acting at the end of the 1970s: changes in firm size, the disappearance of some heavy-industry sectors, economic growth in district areas and the decline of large firms. The argument was that, although composed of SMEs, industrial districts could obtain similar, or even better, results compared to larger firms.

Italian authors, such as Becattini (1992), explained the role of district firms through the decentralization of production or flexible specialization that springs from changes in the environment and production systems. Becattini (1992) studied industrial districts using a territorial area as the unit of analysis and applied Marshall's ideas on production processes to introduce sociological and behavioural variables. He defined the industrial district[8] as a network of external economies and diseconomies, linking costs, historical and cultural factors that determine business and personal relationships. Other Italian researchers that followed along these lines investigated the determinants of the growth of industrial districts always based on the identification of external economies (Bellandi, 2002).

The study of industrial districts and clusters has continued to grow due to the interest of researchers in using systems of firms as units of analysis, the economic development of district areas and the fact that in Italy the industrial districts were recognized as worthy objects of public policy support[9] (Sforzi, 2002; Toni and Nassimbeni, 2003). With increased multidisciplinarity, the industrial district concept became more complex and lost some clarity. Box 2.2 compiles some of the most cited definitions of 'industrial district' which allow a distinction between industrial districts and other geographical concentrations of firms, as well as between the economic and sociological perspectives on industrial districts.

Box 2.2 The concept of 'industrial district'

Production system...

- A production system, geographically localized and based on the division of labour between SMEs specialized in various stages of the same industrial sector (Marshall, 1919).
- An economic system, locally coordinated and well articulated, usually specialized in a single product, characterized by the division of labour among firms, which obtain economies and other advantages from spatial agglomeration (Amin and Robins, 1992).
- A production system, geographically defined, characterized by a large number of firms, mostly small firms or microfirms, which take up various stages of the production of a homogeneous product (Pyke and Sengerberger, 1992).
- An industrial activity area, spatially delimited and with a distinctive specialization with respect to resources, products or services (Markusen, 1996).

...with sociological and cooperative elements

- A reality, characterized by a geographically delimited area centred around a dominant production type; a large number of small firms and microfirms; a flexible organization of production that serves differentiated demand; nonhierarchical, flexible and variable subcontracting relationships; a balance between competition and cooperation, which allows the preservation of the system; and an intense and deep interrelationship between the economic and non-economic (sociological, cultural and ethical) factors (Becattini, 1979).
- A group of firms, located in a small geographical area, which work directly or indirectly for the same final market; share the same values and knowledge; are linked by particular relationships in a complex combination of competition and cooperation (Brusco, 1992).
- A socioterritorial entity, characterized by the internal interaction of a system of small manufacturing firms, which intervene in various stages of the same production process, spatially concentrated, closely linked to local production and sharing a limited territory (Sforzi, 1990).

In the first group of definitions (Marshall, 1919; Amin and Robins, 1992; Pyke and Sengerberger, 1992; Markusen, 1996), industrial districts are considered as production systems specialized in a common product or industrial sector, located in a particular geographical area and composed of a large number of firms, with a strong division of labour among them. This first group of definitions does not, however, allow the differentiation between districts and other spatial concentrations of firms such as industrial clusters. The second group of definitions (Becattini, 1979; Brusco, 1992; Sforzi, 1990) adds to the first the sociological and cooperative elements. In this way, they define a set of firms characterized by the relationships established among them, in what Dei Ottati (1988) called a 'common market'. All in all, it can be said that the first is an economic perspective, which shows a concept of 'district' similar to that of 'industrial cluster', based on the division of labour and specialization. On the contrary, the second perspective emphasizes sociological and organizational factors, which distinguish the cluster from the type of industrial district known in the literature as Italian, canonic, orthodox, pure or textbook.

2.2.3 Empirical literature on industrial districts

The empirical literature on the industrial district as a unit of analysis is essentially interdisciplinary (Belloi, 2006) and has traditionally suffered a major shortcoming in the lack of empirical studies. More recently, that shortcoming has been reduced as a growing interest in the topic has established conceptual bases of investigation. The object of investigation has also been broadened by incorporating issues such as R&D policy and strategic issues.

Table 2.4 provides some examples of empirical contributions that are particularly relevant from the country or industry point of view. Among those that investigated only one industry in only one district, Molina (2001a, 2001b) studied the tile industry in Castellón (Spain), while Such (1995) focused on the textile industry in Alcoi-Ontinyent (Spain). Among those that carried out a comparative analysis of several industrial districts within the same industry, Signorini (1994a, 1994b) looked at the textile industry and Rabelloti (1995) compared four sportswear districts (two in Italy and two in Mexico). Other authors compared districts in different industries: Saxenian (1994) for high-tech industries in Silicon Valley (California) and electronics in Route 128 (Massachussets); Soler (2000) for four districts of the Comunidad Valenciana (Spain) specialized in textiles, toys, tiles and furniture. Some studies also looked at all districts in a region or country in order to investigate their importance for the economy of that region or country: Sforzi (1990) for Italy and Ybarra (1991) for the Comunidad Valenciana.

Further to the latter group of studies, it has been found that there are approximately 200 industrial districts in Italy, which account for more than 40 per cent of manufacturing employment and more than 40 per cent of exports (Goglio, 2002). In Spain, Boix and Galletto (2005) found 237 industrial districts, accounting for approximately 47 per cent of manufacturing output.

Table 2.4 Perspectives of analysis of industrial districts

		Number of districts per sector	
		One	*Several*
Number of sectors	*One*	Such (1995) Molina (2001a, 2001b)	Signorini (1994a, 1994b) Rabelloti (1995)
	Several	Soler (2000) Saxenian (1994)	Ybarra (1991) Sforzi (1990)

All in all, by defining the industrial district as the unit of analysis, it has become a distinguishable concept. In particular, the work of Marshall was necessary but not sufficient to explain the territorial concentration of firms. Although the concept of external economies is the basis for the study of the advantages of agglomeration, it still does not explain the initial conditions under which agglomeration may or may not occur. By distinguishing between various forms of territorial agglomeration, in particular industrial districts and clusters, it has become clear that certain social and organizational characteristics are required for an industrial district to be established. The next section will discuss which characteristics those are, how they impact on the specialization of the territory and on the analysis of the economic and social impact of the industrial district on its territory.

2.3 Strategic implications of the formation of industrial districts for European traditional manufacturing

2.3.1 Social and organizational characteristics: people, relationships and firms

The strategic implications of the formation of industrial districts rest on their structural elements – the stable characteristics that affect its firms: production organization of the territory (Pietrobelli and Olarte, 2002); innovation rates (Cainelli and De Liso, 2004); internationalization (Bechetti and Rossi, 2000a, 2000b); and performance (Signorini, 1994a, 1994b). In order to obtain a comprehensive knowledge of the behaviour and performance of district firms, the analysis of industrial districts requires, together with the study of the firms themselves, the investigation of the social relationships in which the firms are embedded.

From a network perspective, the social and organizational characteristics required for the analysis of industrial districts can be classified into three groups: a community of people, a system of relationships and a population of firms (Figure 2.2).

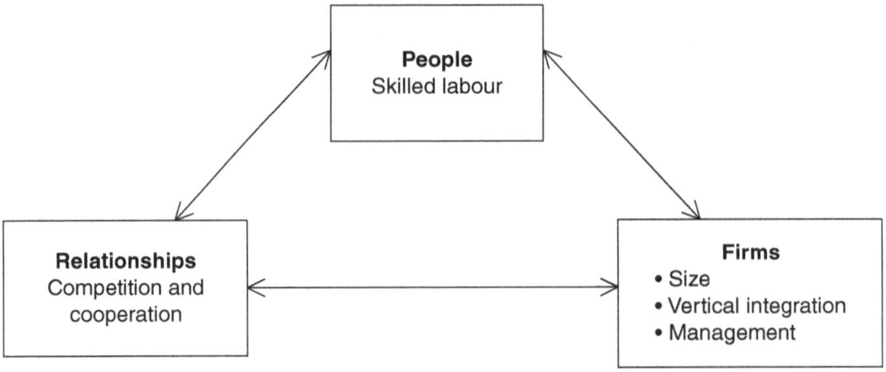

Figure 2.2 The network perspective.

A community of people

The first element implicit in the industrial district is the community of people (workers and managers) that lives and works in the district. Together with firms and the mechanisms that bind them, the community of people defines the system (Dei Ottati, 2002). It is characterized by the competencies of the workforce and the entrepreneurship of the participants, with a relatively homogeneous value system, expressing a long-term view of the importance of networking at work and in the family. This value system promotes entrepreneurship and innovation and guarantees the survival of the district (Becattini, 1992).

Together with the value system, there is a system of institutions (markets, schools, families) that spreads and perpetuates those values across the district (Becattini, 1992). This mechanism creates a feeling of belonging and identifies the inhabitants (Becattini, 1979). From the strong interdependence between people and organizations emerges a social pact of integration that oversees transactions within the district (Dei Ottati, 2002).

The district congregates a specialized workforce that is one of the main sources of external economies. There is solid empirical evidence of its existence and it is characterized by its stability, local recruitment for the most part, high levels of vocational training, and high mobility up the career ladder (Pietrobelli, 2000; Zucchella, 2006).

Together with the local labour market, the district environment favours entrepreneurship (Omiccioli and Quintiliani, 2000). The renovation and growth of the population of firms takes place from within, taking advantage of the flow of resources and business opportunities created by the local market. Besides, those firms that are able to survive the initial stages of life are also able to remain active for longer (Casavola *et al.*, 2000). In this way, a more homogeneous space is formed, with employees taken from other firms that share the same value system, institutions and rules that govern business activity in the district.

A system of relationships

The second important structural element that defines the industrial district is the system of informal and unstructured relationships, which has important strategic implications for district firms. By definition, these firms are embedded in a complex network of relationships. The characteristics of each one and its participation in the value chain, together with the intensity and direction of the interfirm relationships supported by their industry, represent a range of situations and phenomena that have been widely studied.

The literature characterizes such relationships as a combination of competition and cooperation: firms compete against each other, but at the same time they also jointly look for new markets for surplus production (Dei Ottati, 1988, 1994). This specificity differentiates industrial districts from other systems of firms and is made possible by the absence of large firms that might be able to dominate the rest (Becattini, 1992). It further constitutes an organizational principle of the district and one of the local regulating mechanisms that guarantee continuous success (Piore, 1992).

Cooperation in the district involves multiple actors and can be horizontal or vertical. Cooperation between competing firms is defined as horizontal and includes market-sharing agreements, participation in industrial associations, technological institutes and commercial projects that generate mutual benefits. On the other hand, vertical cooperation is carried out by firms at different levels of the value chain due to division of labour and specialization. This type of cooperation can be forward (or downstream), towards client firms producing the final product or carrying out commercial or distribution tasks, or backward (or upstream), towards supplier firms that produce intermediate inputs (Rabelloti, 1995).

The intensity of cooperation relationships, vertical or horizontal, has been explained through geographical proximity, the local characteristics of society and production and the necessity for cooperation induced by technological change (You and Wilkinson, 1994). In this context, many authors explain the success or failure of districts through the quality, development or frequency of those relationships (Saxenian, 1994; Staber, 1997, 1998, 2001; Schmitz, 2000). Some currently analyse the changes that have been occurring in the district relationships (Grandinetti and Passon, 2004; Belussi and Sedita, 2009). The most noticeable characteristic of that evolution is the renaissance of a leading firm that structures and formalizes relationships through its growth strategy. This process creates other types of territorial networks already discussed, such as the hub-and-spoke districts (Sölvell, 2009).

A population of firms

A third aspect that defines an industrial district is the localized process of division of labour and decentralization of production that remains independent of the general market and does not focus on only one or a few firms (Becattini, 1992).

For that to happen, the agglomeration should possess a number of characteristics, such as a large number of SMEs, a low degree of vertical integration, specialization in one or a few stages of the value chain, and the existence of independent firms, with management-ownership, often kept within the family (Bellandi, 2002). In order to increase competitiveness, the agglomeration may include firms from related industries and other institutions such as business associations and technological institutes. To provide an example from the textile industry, an industrial district would include producers of clothing and home textiles, producers of yarn, buttons, dyes and other inputs, producers of textile machinery, financial firms, transport firms, etc.

An industrial district contains a large number of SMEs, some of them micro-firms, grouped in a particular territory, which as a whole can achieve external economies of scale (Dei Ottati, 1988). As an example, in the Comunidad Valenciana in 2000, the average district firm had 22 employees and annual sales worth €2.67 million, while the average non-district firm had 36 employees and annual sales worth €5.36 million[10] (Giner and Santa María, 2002).

District firms have a low degree of vertical integration,[11] with some selling their products directly to the market and others selling intermediate inputs. In this context, Brusco (1992) and Dei Ottati (2002) identify three types of firms in industrial districts: (1) downstream firms, which produce a final product sold directly to the market (in the textile case, clothing or home textiles); (2) upstream firms, which produce intermediate inputs sold to downstream firms (in the textile case, yarn and finished items); and (3) related firms, which offer products and services that complement the main district activity.

There is not a rigid separation between downstream and upstream firms, and the same firm can swap roles in different transactions. According to Becattini (2004), this characteristic defines an operational district, where both types of firms agglomerate. The determination of the optimal degree of vertical integration – the one that maximizes efficiency – is hard to establish, given that the range of possibilities has two opposing structural alternatives as bounds: internalization or externalization of activities.

Camisón and Guía (2001) state that, by searching for higher levels of efficiency, the optimal level of vertical integration depends on the organizational configuration of the industry.[12] In particular, due to the specific structural characteristics of the district, the optimal degree of integration for district firms is lower than for non-district firms. Similar conclusions have been reached by Signorini (1994a, 1994b) for Italian textiles and by Soler (2000) for furniture, footwear and textile districts in the Comunidad Valenciana. However, Soler (ibid.) could not extend the same conclusion to the tile district of the same region, implying that such a characteristic is sector-specific and is mostly linked to production processes with high divisibility (Steinle and Schiele, 2002).

District firms have a particular form of management. First, these are mostly independent firms that seldom belong to large business groups. The drawback of this situation manifests itself in potential limitations to access financial, technological and human resources, and low bargaining power vis-à-vis suppliers and clients that can compromise the growth of such firms. Second, district firms do not typically separate ownership and management and tend to keep both within the family. Third, district firms have a tendency to keep management within the family, which conditions administrative structures: these become simpler, more centralized and less technical.

In sum, from the review of the social and organizational characteristics of industrial districts it can be concluded that there is a strong interdependence between people and organizations, that the relationships between firms are intense and that, on average, the district firms are smaller, have a lower degree of vertical integration (high specialization) and are family firms.

2.3.2 Organization of production: flexible specialization

The adequate combination of the structural elements typical of the industrial district leads to an industrial configuration that explains its evolution and success by means of the decentralization of production and flexible specialization of its firms (Pietrobelli and Olarte, 2002). In fact, SME flexibility – the capacity to respond and adapt to environmental changes – is a key competitiveness factor for the firms (Fiegenbaum and Karnani, 1991) and is at the same time a determinant of success of the territorial concentration system, as the existence of a volatile market favours the agglomeration of industry (Steinle and Schiele, 2002). At the district level, the model of flexible specialization or specialized flexible production is obtained through a process of vertical disintegration, whereby firms specialize in particular tasks vertically linked by division of labour.

The model of flexible specialization has been characterized from both the supply and demand viewpoints (Trullén, 1990). On the supply side, flexibility is a property of the system: following Marshall and Becattini, the social machine is in charge of the fast and efficient adaptation of the district to the evolution of the organization of production and, in this way, controls efficiency at each stage and guarantees the stability of the system. Capecchi (1992), for example, explains that flexible specialization is a production system characterized by a large number of small plants where the organization of labour is based on cooperation, there is a high social mobility of workers and products are custom-made. On the demand side, Piore and Sabel (1984) define flexible specialization as a firm's ability to respond to rapid changes in market demand. This definition supposes that SMEs use flexible specialization as a strategy of anticipation that allows an efficient response to a volatile and diversified demand and justifies its adoption.

From an integrationist perspective, it can be said that flexible specialization springs from the capacity of adaptation of production processes and technologies, requires a large, skilled and available workforce, and is based on a strategy of low vertical integration and a network of social and business relationships that allow the system to respond and adapt to demand shifts.

A related issue is the definition of the sources of specialization and its strategic consequences. For the metallic products district of Gnosjö (Sweden), Wiklund and Karlsson (1994) investigated which types of firms are flexible, in which aspects and how to reach that flexibility. Based on those findings, they proposed the model of flexible specialization illustrated in Figure 2.3, which represents both how to attain flexibility and what the outcome is.

The model describes four different sources of flexibility for industrial district firms that operate through the three basic functions of storage, transformation and distribution: input flexibility (labour market and independent subcontractors), internal flexibility (production processes and qualified workers), output flexibility (product lines and range of clients/markets), and network flexibility (set of relationships with other firms). From the work of Wiklund and Karlsson (1994), it can be concluded that the district flexibility is a mix of four elements, although their optimal combination depends on the technological requirements of the specific industry in which the district is specialized. Moreover, downstream firms are the most flexible as they sell to clients outside the district and are able to adapt to changes in tastes and preferences and to conquer new markets.

Studying the distribution channels used by firms in the textile district of Alcoi-Ontinyent, Such (1995) provides empirical evidence of the higher flexibility of downstream firms. These firms mainly use two distribution channels:

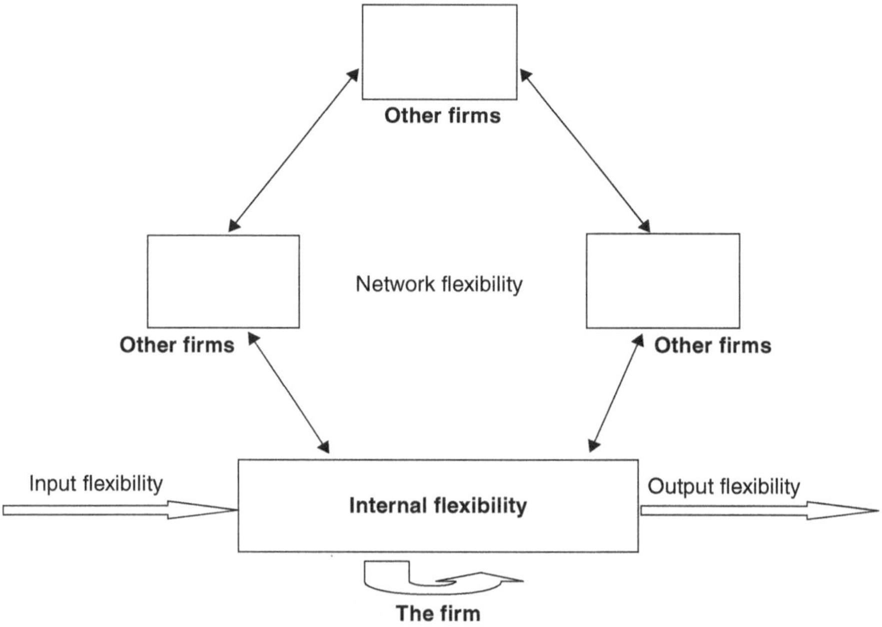

Figure 2.3 The Wiklund–Karlsson firm-flexibility model (source: Wiklund and Karlsson (1994: 109)).

external sales agents and large distribution chains. This is a consequence of the segmentation of textile markets to cover a wider variety of tastes and cultures from a dispersed range of clients. Therefore, as long as the final client does not require the establishment of a close and lasting relationship, and demands a fast and flexible response, one of the main competitiveness factors of the district is the ability to adapt to changes in the business environment.

In summary, the social and organizational characteristics of the industrial district create in its territory a very specialized and flexible industry which, together with the other specificities of the district, leads to differences in performance for district and non-district firms. Those differences can be measured under the concept of 'district effect'.

2.3.3 The district effect on firm performance

A large literature has attempted to measure quantitatively the 'district effect' – the differences in performance between district and non-district firms within the same industry. If the advantages of belonging to a district are realized, district firms would be expected to have a superior performance, measured using balance-sheet data (Devinney *et al.*, 2005).

This line of research provides one possible explanation for the observation that firms have different levels of profitability (Camisón, 2004), although the results obtained are highly heterogeneous (Harrison, 1994). This is due to a number of factors:

- choice of explanatory variables, which is influenced by the field of the researcher (in particular, the determinants of competitiveness can relate to the size, governance and specialization of firms, and performance is not equally related to agglomeration for all firms in the same area);
- use of different levels of data aggregation (regional data or firm-level data);
- adoption of different definitions of 'district' by considering different criteria; and
- the absence of a precise methodology for the identification of industrial districts, which leads to the use of arbitrary thresholds for industrial specialization and region definition (Brenner, 2003; O'Donoghne and Gleave, 2004), with consequences for the validation of the existence of external economies of agglomeration (Soler, 2000) and ultimately of the district location.

Nevertheless, the work of Signorini (1994a, 1994b) on the Italian textile districts of Prato and Biella is a basic reference in the measurement of the 'district effect'. This work measures firm performance through profitability ratios. The database used was the 'Central de Balances' of Italy, from which a sample of 500 textile firms was extracted, with two-thirds of them being district firms. The explanatory variables used were: the degree of specialization, firm size, labour costs, productivity, debt rates and costs. The study found that district firms are more specialized, have higher operational profits, pay higher interest but also have higher debt ratios. This last result is interpreted as meaning that district firms have easier access to credit than non-district SMEs. Subsequent research has reached similar conclusions (Fabiani *et al.*, 1998; Soler, 2000; Puig *et al.*, 2009).

Tattara (2001) has criticized the approach initially proposed by Signorini from the point of view of the methodology and the use of balance-sheet data that may not be applicable to microfirms. The main methodological criticism goes against the comparison between district and non-district firms, as Tattara believed that no good firm operates without a network (firms with which it competes or cooperates) and he favoured qualitative to quantitative methods. The first critique implies some confusion between the general concept of a network and the more narrowly defined concept of a district: even if all firms have their networks, only some firms are part of a true industrial district, and this aspect was neglected by Tattara. The second critique originates in Tattara's belief that balance-sheet data for Italian SMEs are not reliable.[13] However, it is difficult to pinpoint whether this lack of data reliability is specific to Italy or if it also exists in other countries. Besides, poor-quality data are also an issue for developing countries, and it has not prevented quantitative research in that context. Researchers should, of course, be aware of the caveats introduced by poor-quality data and ponder their implications, but it is not a sufficient reason to ignore the use of quantitative research.

Despite the caveats, the district effect methodology that compares district and non-district firms within the same sector has allowed new insights on the performance of district firms. Using firm-level data, a larger number of nominal and ordinal variables can be incorporated, and it is straightforward to isolate firms of different sizes so the robustness of results with respect to the inclusion of microfirms with less reliable data can be checked. Some evidence of the district effect in the Spanish textile-apparel industry is presented by Costa and Viladecans (1999), Soler and Hernández (2001) and Pla and Puig (2009). The first study finds a positive district effect on export intensity. The second provides some empirical evidence for that effect on firms' efficiency and profitability. The third validates the superior international performance of textile firms located in the district of Alcoi-Ontinyent.

It can be concluded that the social and organizational characteristics of the industrial district as a territorial system differentiate it from other models and directly influence the activity of its firms. That influence acts through specialization and externalization of parts of the value chain, and through geographical proximity and frequency of contacts, generating a common business environment that brings together competition and cooperation. The result of this common strategy is visible in the performance of district firms and is measured as the 'district effect'. There is also enough support indicating that district location may influence managerial perceptions.

2.4 Conclusions

Organizational and strategic characteristics change with the environment, so the activities of firms are more and more embedded in networks or systems of complex relationships with a wide variety of actors. Those relationships are built through externalization of activities and the search for flexibility, which allows the firm to adapt in the best possible way to the environmental changes.

Taking into account those changes, the network model has been an example of adaptation to the environment and a reference to evolution. It is a business system that is path-dependent and has been studied from various perspectives. The origin and content of network relationships can be seen as organizational, social and strategic. According to the focus of those perspectives (form, relationships and strategic implications), the network can be classified into different types.

The strategic perspective relates firm size to the degree of specialization and the network power system, as well as relating the firm's organizational characteristics to its performance. It allows an understanding of the firm's strategic spaces that provides a benchmark to study a particular type of network, such as industrial districts.

The literature on single-sector territorial agglomerations explains the external economies and competitive advantages brought about by agglomeration. The analysis of structural and organizational characteristics, as well as the characteristics of firms and their relationships, allows an understanding of the direction and magnitude of their influence on firm performance. However, empirical studies on the strategic influence of location have not generated a consensus with regards to the variables used, hypotheses incorporated or methodology used.

In this research context, and taking into account the environmental changes that firms face, it is important to improve our knowledge of industrial districts, in particular the less well-studied aspects, such as the analysis of their influence on managerial perceptions.

3 Globalization, internationalization strategy and industrial districts

The process of globalization has been defined as 'the increasing economic integration of the world economy' (Deardorff and Stern, 2002) or 'the increasing internationalization of the production, distribution, and marketing of goods and services' (Harris, 1993). Although the concept of globalization involves a range of aspects that are economic, political, sociological and environmental (Fischer, 2003), here the focus will be on the economic dimension of globalization, defined as a process of integration and internationalization.

In modern times, the Industrial Revolution and the post-Second World War era marked two important stages of globalization, which also brought growth and convergence to the world economy (Williamson, 1996, 1997, 1998). The role of globalization in promoting world income convergence through the continuous reduction in transport costs was demonstrated theoretically by Krugman and Venables (1995). Since the 1990s, the globalization–growth nexus seems to be driven by the integration of emerging markets (Brazil, Russia, India, China – the BRIC countries) into the world economy and by the internationalization of their firms.

The liberalization of international trade has led to an increase in the volume of trade, which has been due more to the growth of the 'extensive' margin (greater number of products traded) than the 'intensive' margin (higher quantity of the same traded products). Most of this trade is based on the fragmentation of production that allows an increasing number of SMEs to participate in international markets. Moreover, the rapid pace of innovation in communication and transport allowed some non-tradable services to be traded, creating new markets and new opportunities (Cairncross, 1997). In addition, the pressure by consumers and competitors to innovate, implying high costs of R&D, leads firms to actively enter new markets or find more advantageous locations with lower production costs. Foreign direct investment (FDI) has contributed significantly to the development of some emerging markets, transferring technology from more advanced to less advanced countries and allowing the vertical integration of production.

The process of internationalization of firms has been defined as: the process of growing commitment to international markets (Welch and Luostarien, 1988); the process whereby firms increase their knowledge of the importance of business transactions for their future, as well as their knowledge of the establishment and patterns of transactions with other countries (Beamish, 1990); and the process whereby firms increase their commitment to international trade (Fletcher, 2001). The first of these definitions, proposed by Welch and Luostarien (1988), is perhaps the most comprehensive and understandably the most used. It views internationalization as a dynamic and evolutionary process that is not exclusive to multinational firms, and the relationships established during that process as influential factors in the growth and expansion of the firm to other countries.

In the current economic environment, the globalization of the economy and the internationalization strategy of the firms are two intertwined aspects that affect survival of district firms (Zucchela, 2006; Belussi and Sedita, 2009). Globalization brings a range of difficult circumstances for district firms, and to them the strategy of internationalization seems to be the best response in order to maintain competitiveness.

One of the most visible and threatening manifestations of globalization in industrial districts is delocation as an internationalization strategy. Delocation is seen here in a broad way as the process whereby firms source or transfer part or all of their activities (especially the most labour-intensive ones) or assets to other locations outside their national borders. This can happen by importing inputs, subcontracting internationally or shifting production facilities abroad, and it is generally justified by an attempt to reduce costs. For that reason, delocation typically happens towards lower-income countries, such as the new EU member countries in Central and Eastern Europe, EU candidate countries in Southeast Europe, or emerging markets in Asia and Latin America (Myro and Fernández-Otheo, 2004; Oxelheim and Ghauri, 2004; Denis *et al.*, 2006).[1] By the same token, delocation is a phenomenon most pervasive in traditional manufacturing.

Among other impacts, delocation forces important changes to the established organizational model and facilitates knowledge spillovers to other production systems, in this way weakening the industrial district. On the other hand, the global orientation of the industrial district can influence in various ways the internationalization strategy of SMEs, either stimulating access to external markets through a favourable social environment, or inhibiting the adoption of delocation strategies due to the strong interrelationship between firms and local communities that originates a 'shared strategy'.

The district location influences business activity and internationalization patterns, although the advantages of territorial networks can be reduced by environmental changes. In this context, it is important to understand how globalization is affecting industrial districts and how these are influencing the internationalization strategy of their firms.

Following the sequence illustrated in Figure 3.1, this chapter analyses the impact of globalization on industrial districts and studies the effect of that organizational model on its firms. The first section reviews the concept and evolution of delocation. The second section analyses the impact of globalization on industrial districts and the ways in which they respond to it. The third section investigates the effect of the district location on the internationalization of district firms.

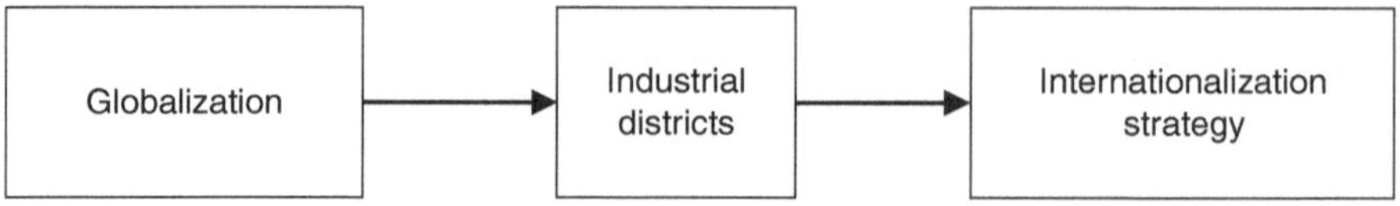

Figure 3.1 Industrial districts as moderators of the globalization–internationalization relationship.

3.1 Industrial delocation: concept and evolution

Firms in high-income countries have been making important changes to their relationship with suppliers and clients in other countries during the last three decades. That evolution has been visible in the reorganization of production, particularly in traditional manufacturing, and can be analysed through various international trade indicators, such as export and import intensity, or the share of imported inputs in output value.

As highlighted by Rodrik (1998), trade and investment are two of the main channels through which globalization impacts upon emerging markets. A very important feature of the current globalization stage that brings together trade and investment and has attracted increasing attention is the phenomenon of outsourcing. Feenstra and Hanson (1996) define it as 'the import of intermediate inputs by domestic firms'. It stems naturally from the international fragmentation of production, which has important implications for wages and employment in developed and developing countries. Feenstra (1998) provides a review of the main issues raised by the process of fragmentation of production, the most contentious of these being the impact on employment and wages of low-skilled workers.

While outsourcing has long been a research topic in international economics (Helpman, 2006), the business economics literature has mainly focused on export markets, looking very marginally at foreign suppliers (Dana *et al.*, 2007). This state of affairs raises problems for three reasons. First, the success of exporters depends on their behaviour in supplier markets (Overby and Servais, 2005). Second, international sourcing is a common practice that has been implemented by 40 per cent of the firms (UNCTAD, 2004). Finally, international sourcing creates situations in which the reorganization of production leads to the partial or total transfer of activities and assets to other countries (delocation). This practice may erode the competitive advantage associated with a territorial form of organization of production.

The industrial delocation decision is complex and ambiguous. On one hand, it is not always carried out simply to reduce labour costs, but also hides structural weaknesses of the territory. On the other hand, it can take several forms, such as imports, subcontracting or relocation. It is important to distinguish national from international reorganization (Table 3.1), which cause national or international subcontracting or outsourcing. A second important distinction relates to firm ownership. A delocation that occurs internationally but within the firm, by

Table 3.1 Firm reorganization models

		Firm	
		Internal	*External*
	National	Relocation of assets (does not require delocation)	National subcontracting or outsourcing (sourcing to domestic firms)
Territory	*International*	Tied delocation (between foreign subsidiaries) Classical delocation (effective transfer of assets)	Imports or international outsourcing – externalized delocation (sourcing to foreign firms)

moving activities to foreign subsidiaries, can be called 'tied delocation'. Alternatively, if it happens outside the ownership limits, either by buying intermediate inputs or subcontracting activities previously carried out within the firm, it is called externalized delocation. Finally, 'classical delocation' takes place simultaneously at the international level and by transferring assets abroad (investing).

Moreover, it is also the case that many firms carry out several kinds of delocation simultaneously. This combined strategy of internationalization, known as multilocation, is made possible by the simultaneous participation of very large firms in different markets, so that for them it is optimal to keep a low level of commitment and control in some markets and higher levels in others. In this way, the same firm may opt to import and subcontract in one market and to carry out FDI in another.

In sum, four types of delocation can be considered as a result of vertical disintegration of business activity carried out at the international level: (1) classical delocation, where plants are closed domestically and reopened abroad; (2) externalized delocation, established as a contract between firms from different countries and ownerships; (3) tied delocation, based on the transfer of activities between subsidiaries of the same firm located in different countries; and (4) multilocation.

The analysis of delocation at the firm level is not straightforward, as the decision is internal to the firm and is often confidential. It is typically made public only in the recipient country and materializes in the transfer of assets from the origin to the destination country. Classical delocations are the ones with a more painful impact on the origin location, as they result in economic and social costs due to the loss of direct and indirect jobs, political costs arising from sunk costs of previous government investment and costs associated with the loss of competitiveness of the territory.

Although the delocation strategy carries short-run adjustment costs with economic and social implications, from the economic point of view it is not a zero-sum game. In the long-run, the international relocation of production increases social welfare: it benefits consumers through lower prices and firms through higher profits. Any jobs lost in the origin country are transferred to the destination countries, which increase employment and income levels. In the origin countries, those workers who lost their jobs are, over time, reabsorbed into other occupations, although how long the process takes and what the reabsorption rate actually is remains a matter of discussion. Moreover, although wages paid by foreign firms in developing countries are typically higher than those paid by domestic firms, reports from the United Nations or the International Labour Organization have highlighted the precarious working conditions the workers are often subject to.

3.2 Industrial districts and the new economic environment

During the last three decades, the changes induced by the opening up of world markets – in particular of capital markets that finance investment worldwide – and by technological progress – which allowed improvements in transport and communications, with steady and significant reductions in their costs – have forced the reorganization of production systems. The role of strategies of innovation, decentralization of production, outsourcing and internationalization has been reinforced.[2] At the same time, globalization promoted the convergence of tastes and demand patterns across the world, brought about by increased knowledge of other cultures and the action of the mass media; the boom of regionalism with the creation of free trade areas across the world; the consolidation through national and international mergers and acquisitions (particularly in the EU around 1992, when the Single Market was launched); and the 'delocation' of production in areas that are a source of competitive advantages for independent firms and subsidiaries.

Firms have developed strategies of internationalization to respond to the environment they currently face. Julien (1996) offers a classification of actions employed by SMEs in high-income countries which combines two dimensions (Figure 3.2). Inputs used in the production process can be sourced locally or imported, and the destination markets for the output can be domestic or international.

The classification in Figure 3.2 shows six types of SMEs:

1 Those that are not affected by globalization due to protection that keeps them safe from competition.
2 The SMEs that are present only locally, sometimes as producers of intermediate inputs or producing for local markets.
3 Those firms that resort to international markets as suppliers of intermediate inputs and services.
4 The SMEs that source locally and export a certain share of their output.[3] Within this group three types of firm can be considered: (1) 'opportunists', who export occasionally when an order is placed from abroad; (2) 'professionals', who continuously export a high percentage of their output and reach far-away markets; (3) 'hesitants', who hesitate between a greater commitment to or abandonment of the international strategy altogether, due to the growing effort required by this activity, which can have a negative impact on the development of their local strategies.

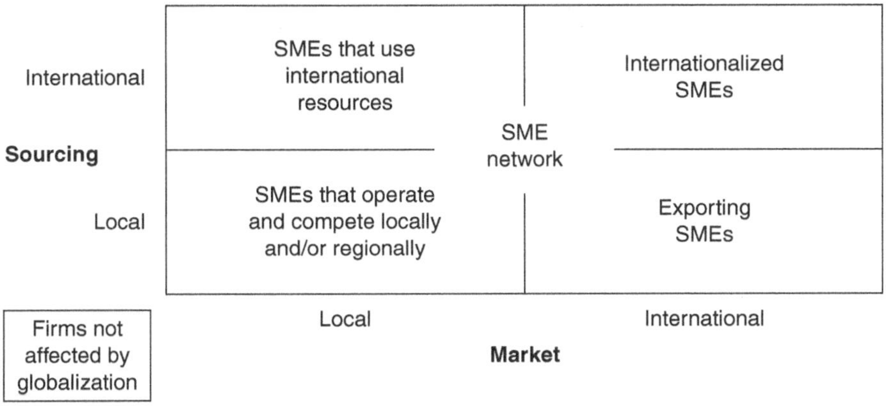

Figure 3.2 Typology of responses of SMEs to globalization (source: Julien (1996)).

5 Internationalized SMEs, which are regularly present in foreign markets, both sourcing intermediate inputs and services and selling their output, in many cases using complex forms of international participation, such as contractual systems and FDI.
6 The SME networks, which internationalize collectively, such as SMEs in industrial districts. In those systems, there may be firms in any of the four quadrants of Figure 3.2, with some competing locally, others internationally, and others using only sourcing or only exports.

Although this classification allows an observation of the different types of firms' responses to globalization, it does not satisfactorily explain the factors that hinder and promote the adoption of the internationalization strategy and the process followed by firms when they internationalize (Fletcher, 2001). From this point of view, it is important to know, with respect to internationalization, why firms in the same country and the same sector take different levels of risk in foreign operations and, with respect to the territory, what the consequences of that kind of behaviour on the production organization of industrial districts can be. These questions are relevant both because, as shown previously, internationalization is a key strategy to competing successfully in a globalization context, and because the internationalization of district firms, particularly on the sourcing side, is an important challenge for their survival.

The literature points out factors that hamper the internationalization of SMEs:

1 Financial limitations due to lack of access to credit and to hedging against exchange rate fluctuations.
2 Commercial difficulties in identifying opportunities in foreign markets, selecting reliable distributors or other means of distribution and supply of exported goods.

3 Cultural issues in management (prejudices, lack of knowledge about foreign cultures, interpretation of foreign legislation, red tape issues).
4 The small size of the firms and lack of skilled and experienced staff.
5 The existence of strong foreign competitors, especially from emerging markets.

On the other hand, the literature suggests that the main factors helping the internationalization of SMEs, and with it the establishment of a key strategy in the globalization environment, are:

1 The growth potential of new markets, especially after the establishment of free trade agreements.
2 Diversification of risk by operating in several markets.
3 Excess capacity in existing markets.
4 Imitation of the strategies of successful firms.
5 Achieving economies of scale.
6 Belief in external markets as natural markets, given that firms are created to operate internationally.
7 Higher profitability of international markets where products may benefit from higher preference and command higher prices.
8 A strategy of serving clients at their location.
9 Taking advantage of tax incentives and cheaper labour.
10 Escaping the pressure of local competition.
11 Following important clients that have also internationalized.
12 Access to technology and raw materials.

These determining factors act on the responses of SMEs in a way that can be moderated by the territory in which the SMEs locate. The impact of the internationalization strategy can be measured through an analysis of the flows of exchange (rates of delocation) and their content (reorganization models). The measurement of impacts is of particular interest to the study of industrial districts as it allows an assessment of whether delocation is a threat to or an opportunity for their survival (Zucchella, 2006).

Accordingly, Mariotti (2004) argues that the delocation of production can be a threat to the district model as it harms the social context by increasing firms' mortality ratios and reducing the number of jobs, changes the organization of production, value systems and interfirm relationships, and reduces competitiveness due to the increased transfer of specialized knowledge and the externalization of strategic productive competencies internal to the district model. On the other hand, Mariotti (ibid.) also suggests that delocation can be seen as an opportunity if it fosters the local context through the generation of new production systems linked to the original industrial districts and if it contributes to the competitiveness of the district through a differentiation of specialization and competition between countries.

Nevertheless, that assessment does not take into account aspects related to institutional and technological change that may accelerate or delay the impact of delocation. For this reason, it still remains to be understood what the current scenario faced by industrial districts is and how they respond to globalization, or how globalization impacts on their decisions, especially from the point of view of the delocation of their manufacturing activities.

3.2.1 Delocation and industrial districts

Delocation is a strategic response of industrial districts to globalization that can be studied from several angles: why to delocate, what to delocate (a production stage, a production line, or the whole plant), when to delocate, where to delocate from and where to delocate to, and how to delocate (Table 3.2).

Why to delocate in an industrial district?

In a globalized economy, delocation is a competitive strategy implemented by firms in order to gain a number of competitive advantages that would be difficult to obtain otherwise (Porter, 1998). This is also true in the context of industrial districts (Basevi and Ottaviano, 2001).

The competitive strategies defined in this way are offensive or defensive actions that firms pursue due to external changes in markets and internal changes in innovation capabilities. When those actions place firms in a privileged competitive position that allows them to obtain above-average revenues, they are considered as opportunities that can be converted into competitive advantages. This justifies that, in spite of the advantages of the district location, some firms transfer their assets to other territories with even more favourable conditions.

In other cases, the internationalization strategy can be explained by an unfavourable evolution in the conditions faced by the production factors and the necessary redefinition of the firm's strategic behaviour. In this way, factors such as the decrease in transport costs, the deregulation of international trade in the context of the WTO, lower labour costs or tax benefits open the door to future competition by new producers.

Crestanello and Della Libera (2003) studied the causes of the transfer of production operations by firms in the fashion district of Vicenza (Italy) to Romania. They found that the two main motives for the delocation had been the reduced size of the labour force available at the original location and the long labour-intensive value chains that characterize the activity.

What to delocate in an industrial district?

Prior to analysing the object of a total or partial delocation, it is necessary to define location as the place where production factors (capital, land, labour and technology) are transformed into products or services for sale to other firms or final consumers.

Table 3.2 Responses of industrial districts to globalization

Delocation	*Theoretical background*	*Conclusions*
Why	Competitive advantage	The main causes of the delocation of district firms are labour scarcity associated with high labour intensity. Example: Vicenza fashion district (Crestanello and Dalla Libera, 2003).
What	Location theory	The delocated production activities are the most divisible, with low value added and are more labour intensive. Example: Udine furniture district (Grandinetti *et al.*, 2004).
When	Life cycle	Delocation becomes more pervasive when the product becomes more standard and innovation stops. Example: Brenta footwear district (Rabellotti, 2003).
Where from/ where to	External economies	Geographical specialization can act as an inhibitor of the delocation decision. Less strategic activities are delocated to regions that offer cost advantages. Example: Italian footwear districts (Amighini and Rabellotti, 2006).
How	Eclectic paradigm of FDI	District firms are, on average, smaller and more embedded in the social context, hence their delocation strategies imply a lower resource commitment and lower levels of outsourcing. Example: 'Made in Italy' districts (Mariotti, 2004).

In the tradition of Weber (1910), German location theory explains that the location choice for a plant depends on several aspects related to the cost of production factors and their mobility costs, including barriers to their circulation. For those reasons, natural-resource-based firms need to locate close to the source of raw materials. It is similar for district firms, who draw their competitiveness from the network in which they are integrated, or for firms that require proximity to and knowledge of the markets.

In this context, when talking about what to delocate it is important to clarify that when a firm depends on an exclusive resource such as specialized management, bargaining power or innovation, it is unlikely to delocate. It may do so partially for those production stages that are less dependent on location, less profitable and more divisible, keeping at headquarters only those stages in which it is more competitive.

In those industrial districts centred in an activity with low profitability that is divisible into production stages, delocation will be more frequent. For example, textile yarn obtained from wool, cotton, or synthetic fibres can be processed independently and those raw materials are easily obtainable anywhere. With the possibilities opened up by globalization, some European district firms specialized in the first stages of production have been delocating to Central and Eastern European countries, which offer lower labour costs and adequate raw materials (Mariotti, 2004; Zourek, 2007).

Moreover, firms within each sector differ in the structure of the value chain and in the extent of external economies obtained at their location. This is important if considering that, when formulating its delocation strategy, the firm does not consider in isolation the production activities to be delocated. Rather, it takes into account the various interrelationships between all its activities and the value system composed of suppliers, clients, competitors and administration.

Consequently, a large firm that is vertically integrated is less dependent on its specific environment and has enough capability to delocate abroad. Its competitiveness is based on the idea of cross-subsidization of functions: more efficiency in some functions (for example, lower production costs) can reduce costs in other functions. On the contrary, a specialized SME competes mainly on flexibility and depends more on the coordination of tasks than on each particular task, so it depends mostly on relationships established with the environment where it is located. In this context, Grandinetti *et al.* (2004) studied four leading firms of the furniture district of Udine (Italy) and found that those activities delocalized through the acquisition of plants in Eastern Europe have low value added, are 'decomposable' and labour-intensive.

When to delocate in an industrial district?

Globalization, market deregulation and advancements in information and communication technologies are necessary, but not sufficient, conditions for the surge in delocation. Other factors that are country- and industry-specific also determine the outcome, and for that reason delocation has not taken place at the same time or to the same extent across developed countries. European countries such as France, Italy and Spain have more heavily regulated economies, with more rigid labour markets, making firm relocation more difficult and onerous (Hadjikhani and Ghauri, 2001; Guercini, 2004).

On the other hand, the product lifecycle theory (Vernon, 1966) states that when a new product is introduced, its production will locate close to where the innovation took place, and the product will be distributed nationally and internationally (exported) from that location. As production advantages are lost and the product is standardized, firms relocate production to foreign locations with more favourable labour costs and tax rates. Over time, they may totally abandon the initial production location (Belussi and Sedita, 2009).

The product lifecycle can explain why the delocation rate in the telecommunications sector has been the highest in its pioneering countries, such as the United Kingdom and the United States, as firms from those countries have found favourable conditions for relocation to emerging markets, such as India. A similar behaviour can be observed in the case of suppliers of vehicle components (Autotex, Hewlett Packard, Lear, Valeo), which have delocated to the new EU members as these achieved very high growth rates with the transition to fully functioning market economies. Another example comes from the footwear district of Brenta (Italy), studied by Rabellotti (2003). Comparing firms in average and luxury market segments, he found that firms in the more innovative segment (luxury) are less dependent on foreign suppliers and delocate a smaller number of processes.

Where to delocate from and where to delocate to in an industrial district?

The reasons for firms to leave particular locations in favour of others can be grouped around two major factors: comparative advantages and external economies of agglomeration. These factors have been well documented in the economic geography literature (Fujita *et al.*, 2000; Baldwin *et al.*, 2003) and can be traced back to authors such as David Ricardo and Alfred Marshall. The comparative advantages of a region provide the initial conditions that explain the spatial distribution of economic activity and can act as push and pull factors. Firms tend to locate the various production stages and functions according to the availability of production factors such as labour, land, capital and technology. Hence factor availability will determine specialization and the international allocation of production stages and tasks. On the other hand, entry and exit barriers and economies of scale are also important determinants of the geographical distribution of firms. In turn, the geographical location and concentration of firms in particular sectors across different countries shapes the patterns of trade and specialization of those countries (Fujita *et al.*, 2000).

Agglomerated firms in a given sector enjoy strategic advantages that allow them to adopt different patterns of delocation. Crestanello and Della Libera (2003) and Amighini and Rabellotti (2006) argue that the various specialization patterns inherent in a certain activity and territory influence the competitiveness of its firms and in this way act as moderators and inhibitors of the delocation decision. In particular, they observe that some Italian footwear districts that specialize in low-price market segments have outsourced a high share of their production to Southeast Europe, while other districts that specialize in higher-value segments tend to outsource production to a lesser extent and sometimes to other EU-15 countries.

How to delocate in an industrial district?

The delocation strategy has traditionally been analysed by means of an eclectic approach (Dunning, 1981, 1988, 1992). According to this tradition, firms will decide to transfer production to other countries depending on a number of conditions: the existence of advantages internal to the firm that allow it to compete internationally; the internalization economies that make it more advantageous to keep production internal to the firm than to outsource it to other firms; the higher profits from locating production activities in other countries; and the consistency between the firm's international activities and the firm's long-term strategy.

Mariotti (2004) observed that Italian district firms did not meet that set of conditions as much as non-district firms. Moreover, social interaction in the district decreased the opportunism of management. As a result, district firms preferred indirect investment under contracts, while non-district firms, larger and less socially bound, preferred direct investment under ownership.

In conclusion, and in line with the arguments of Zucchella (2006) and Sammarra and Belussi (2006), it can be expected that the delocation strategy of industrial districts in a globalization environment will be inhibited by a number of aspects that characterize and condition the activity of district firms. Responding to a strategic survival need, the global reorientation of the district from its local nature favours the relocation of those activities (or value chain tasks) in which district firms are less competitive towards other firms located also within or outside the district. This process is generally called externalization. It is also possible to argue that some activities are better suited for externalization than others. Activities with higher value added or that are more central to the territory's development strategy are generally less prone to externalization. Moreover, the social and organizational characteristics of the district based on mutual trust dampen the adoption of models of international reorganization of business activities, especially those that demand a higher commitment of resources.

3.3 The location effect in the internationalization process

Although there is some consensus in the classification of the strategic responses of the SMEs with respect to globalization (Julien, 1996), there are still important gaps in the understanding of their internationalization process and, in particular, regarding their multilocation operations. These are important questions because, in the current environment, the internationalization strategy is a key aspect of the survival of manufacturing firms, and in this respect the behaviour and results observed for district firms greatly differ from those observed for non-district firms.

The internationalization process can be studied from both an economic and a behavioural point of view (Elo, 2004). Whereas the former identifies the determinants of the process, the latter focuses on the stages of development of the process. Within the behavioural perspective two strands can be further distinguished: the study of aspects internal to the firm (Uppsala Model and Innovation Model) and the study of the social context of the firm (Network Model). Both strands take the view that the internationalization of the firm is incremental due to the lack of knowledge about external markets and the uncertainty associated with the internationalization decision.

3.3.1 The Uppsala Model and the Innovation Model

The Uppsala Model (Johanson and Vahlne, 1990, 2009), also known as the U-Model, postulates that the internationalization of firms follows a sequential process, dependent upon the gradual acquisition of the knowledge required to face the unknowns of foreign markets. Hence, a firm's international growth is constrained by the uncertainty arising from foreign operations, and it is the accumulation of knowledge capital on foreign markets that permits increments in the levels of international commitment.

Accordingly, firms start their international activity by exporting, sometimes only occasionally or through agents, and progressively commit to more complex international activities, such as the establishment of sales affiliates, and later invest directly in export markets through production subsidiaries. Moreover, international activities will first be carried out in neighbouring countries, where cultural and historical affinities diminish the psychological costs of doing business. Over time, firms will progressively extend their activities to other countries located further away geographically and culturally.

A similar argument for the establishment of trade and investment links between countries in close proximity is provided by the empirical studies using gravity models (see Helpman and Krugman (1985) for a theoretical basis and Spies and Marques (2009) for a link between theory and empirics). A similar view on the importance of cultural proximity is taken by Hofstede (2001). Generally, geographical distance is strongly correlated with cultural distance for a number of reasons, not least historical and linguistic ties fostered over centuries between neighbouring countries. A good example is the rapid increase in trade and investment links between Portugal and Spain since 1986, when both countries became EU members and the policy barriers to trade and investment were removed.

The U-Model has met with wide support due to its applicability to all types of firms, and it has been particularly used to study the process of internationalization of SMEs. Firms learn from expansion abroad, and knowledge spills over to other firms in the same country, fostering a strong correlation between international experience and the utilization of more advanced forms of internationalization. However, the U-Model does not explain the reasons that lead firms to begin internationalization. Its explanatory power has become more limited in recent years due to the rapid changes in business environments that allow firms to jump over internationalization stages and reduce internationalization lags to the point of eliminating them, as is the case with the so-called 'born global' firms (Aspelund and Moen, 2005).

In an attempt to overcome the previous critique, Johanson and Vahlne (1990, 2009) tried to explain internationalization from the point of view of the permanent interaction between status (knowledge of markets) and change (commitment levels). They argue that the internationalization process is influenced by the interaction between knowledge and commitment. A better knowledge of foreign markets will lead firms to increase their commitment levels to those markets, which in turn will allow them to accumulate more knowledge and again induce a transfer of resources. Johanson and Vahlne recognize, however, that the sequential process can be affected when firms have enough resources to jump stages, market conditions are stable and firms have experience in similar markets.

Another theory that tries to explain the process of internationalization is the Innovation Model, also known as the I-Model. This model, first proposed by Lee and Brash (1978), is based on the application of the principles of diffusion and development of innovation to the internationalization decision. According to the I-Model, the internationalization decision is rationally planned and depends on the information available to management. In this way, the extension of commercial activities to foreign markets could be understood as the result of a process of acquisition of knowledge (information) generated by other firms that have already internationalized.

Bilkey and Tesar (1977) suggested that the sequential process of export development that constituted internationalization goes through six stages. In the first stage, the firm was not interested in exporting, while in the last stage the firm might be exporting to more distant countries. Each new stage represents a greater involvement with international activities and so it is a function of the accumulation of knowledge required to reduce the uncertainty associated with the internationalization decision.

Cavusgil (1980) proposes an extension of the original internationalization model with five stages. In the first stage of domestic marketing, the firm is not interested in exporting and so it sells its products only in the domestic market. In the second stage of pre-exporting, the firm's management realizes the opportunity to export, becomes interested in it and searches for information. When the firm starts taking occasional orders from neighbouring markets, it enters the third stage of experimental involvement. The commitment to foreign markets is very fragile. The firm experiments with certain products and, in doing so, it acquires knowledge about the performance of its products abroad. In the fourth stage of active involvement, the expansion strategy towards foreign markets includes exporting a higher share of production and/or a larger number of products to a higher number of and perhaps more distant countries. In the last stage, firms depend to a large extent on foreign markets for revenues and materials. The remaining decision concerns the allocation of limited resources between national and international markets. In this stage of committed involvement, some firms may adopt more advanced methods of internationalization beyond exporting, such as cooperation agreements and FDI.

The sequential processes described so far establish that internationalization is a stage-based process in which the acquisition and accumulation of knowledge play an important role. Still, the models presented do not explain the timing of internationalization, the utilization of more complex forms of internationalization such as FDI, or the geographical organization of industries and the effect of location. Other literature has studied the factors that influence the international expansion of the firm (see Bernard and Jensen (2004) for the United States; Roberts and Tybout (1997) for Colombia; Aitken *et al.* (1997) for Mexico; and Clerides *et al.* (1998) for Colombia, Mexico and Morocco). A consensus has emerged around a number of factors that seem particularly important, such as firm size, type of products offered, industry characteristics, experience (or age of the firm), expectations of managers, and trade policies.[4]

For the Spanish case, the internal decision-making process that leads to internationalization has been studied, taking into account both the U-Model and the I-Model, with divergent results with respect to the firm's age, technological capability, productivity and level of reinvestment. For the textile-clothing industry, Eusebio and Llonch (2006) confirm the positive effect of R&D activities on the start of internationalization and its exporting intensity, but in general there is no consensus regarding the determinants of the process and existing results hint at the importance of institutional and territorial aspects.

In fact, the globalization conditions faced by European traditional manufacturing firms have revolutionized the industrial organization models based on large firms and give support to other models based on interorganizational relationships, networks, SMEs and territories. Hence, it is important to ask how those interrelationships have affected the internationalization processes. For example, the environment surrounding entrepreneurs is an important source of information and resources that influences the opportunities available to firms. Besides, as highlighted previously, networks have important implications for the formulation and performance of strategies. In this way, integration in a network conditions the internationalization processes due to the interaction and trickle-down effects induced by the network itself. Together with the territorial elements of productive flexibility and coordination, the environment will be favourable to international activities. In this context, the need to study different perspectives of internationalization that consider locational issues and networks is justified.

3.3.2 The network model

The network model of Johanson and Mattsson (1988) tries to explain the internationalization of firms and has served as inspiration for more recent work on newly internationalized firms (NIF) and district firms. They highlight the links between the firm and the industry in the internationalization process. Firms develop a number of relationships with clients, distributors and suppliers, creating an industrial market. Those relationships define various types of networks that allow an understanding of the growth of the firm. Hadley and Wilson (2003) argue that the degree of internationalization of the network to which the firm belongs determines the access to information and contacts that help in explaining the internationalization process of each firm.

Johanson and Mattson (1988) define an industrial system as being formed by interconnected firms in the production, distribution and consumption of goods and services, so that the resulting industrial organization is a network of firms. In such a system there is a division of labour that makes firms interdependent, but there is also a need for coordinating activities. However, coordination does not occur through ownership – it follows a market system with a price mechanism (Langlois and Robertson, 1995).

According to the network configuration defined by Johanson and Mattsson, the internationalization of the firm will depend more on the set of relationships that it maintains within the network than on possessing any specific advantage (Coviello and McAuley, 1999). In this way, internationalization can be achieved through three stages, subject to network relationships: in the first stage of extension, the firm establishes relationships with suppliers, competitors and national clients that are already internationalized; in the second stage of penetration, the firm starts committing more resources to international markets; in the last stage of international integration, the firm develops and establishes relationships with a larger number of markets (Johanson and Mattson, 1988).

According to the Network Model, any global industrial sector can be characterized by various criteria linked to geography and the product and process mix. From the geographical point of view, one can distinguish national and local sectors, where the latter includes more intensive relationships that span all of the firm's activities. Similarly, it is possible to distinguish a number of subsectors within one particular sector, depending on the existing product variety. For that reason, in traditional manufacturing such as textile-clothing, those areas where firms specialize in a particular subsector of activity are strongly interconnected to others through relationships of mutual interdependence. That characteristic would explain to some extent the heterogeneity in the participation and exploitation of the territorial advantages and could guide the restructuring strategies needed by the textile-clothing industry.

In conclusion, the internationalization of a network of local producers differs from internationalization at the national or global level. In turn, at the national level there are differences between the levels of internationalization of a production network (territorially localized) and the industry to which they belong. At the global level, there are differences between the levels of internationalization of the same sector across different countries.

In the case of China, Zhao and Zou (2002) do not find any evidence that geographically concentrated firms show a higher propensity to internationalize; in the United States (Mittelstaedt and Ward, 2003) and France (Chevassus-Lozza and Galliano, 2003), a positive relationship between territorial agglomeration and propensity to internationalize has been found. For the case of district networks in Italy, Becchetti and Rossi (2000a, 2000b) found that the propensity to export and the export intensity of district firms is higher than that of non-district firms. The difference is even higher for smaller firms and firms in mature sectors.

Johanson and Mattson (1988) describe different types of firms using the degree of internationalization of firms and markets (Table 3.3). When the degree of internationalization of the clients, suppliers and competitors of a network is low, two types of firm are considered. One type is the initiators, which initiate international activities that demand little commitment of resources. The other type appears when the commitment of resources is higher – these firms are called solitaires. Due to the continuous involvement in foreign markets, those firms have already acquired the information required for a higher level of internationalization and so perceive a lower risk in internationalizing. In the opposite situation, when the degree of internationalization of the clients, suppliers and competitors in a network is high, two other types of firm can be considered. The stragglers still have a low degree of commitment to foreign markets, but due to the high level of information and knowledge of foreign markets provided by the network, it is very likely that the firm will devote an increasing amount of resources to internationalization. Finally, the jointly international firms show high levels of internationalization, together with all other firms in the same market. In this situation, the firm has access to a very high level of information, well above that of firms that have internationalized on their own. As such, it shows high levels of commitment, establishing production affiliates or expanding to a larger number of countries.

There is a strong body of research dedicated to the role of networks as determinants of internationalization. Chetty and Blankenbourg-Holm explain how the characteristics of the firm and the market to which they belong influence the decision to enter foreign markets and the subsequent expansion to other markets (Chetty and Blankenbourg-Holm, 2000a) and show how the characteristics of management influence the firm's response to stimuli from the network, such as export promotion programmes sponsored by national governments (Chetty and Blankenbourg-Holm, 2000b). Most authors defend the Network Model as explaining the internationalization of SMEs better than the U-Model.

Table 3.3 Internationalization and networks

		Degree of internationalization of the market	
		Low	*High*
Degree of internationalization of the firm	*Low*	Initiator	Straggler
	High	Solitaire	Jointly international

Source: Johanson and Mattson (1988).

In fact, the Network Model makes several contributions. First, it complements and develops theories such as the U-Model, but it also abandons the idea that internationalization results from a rational decision process – instead, it is influenced by the network. Second, the Network Model introduces the element of market relationships, which can affect the choice of market and the mode of entry. Finally, the Network Model considers the characteristics of firms and markets jointly. The study of markets is very important in industrial systems characterized by local cooperation and global competition, such as the district model. Still, Chetty and Blankenbourg-Holm (2000b) point out the lack of consideration by the Network Model of the role of management in the whole process, the external stimuli of the government and the dynamics of how firms go from one stage to the next. All these aspects must be taken into account when making strategic and political recommendations.

The role of networks in theories of internationalization has been growing.[5] Of particular interest here is the study of the impact of localization on the internationalization of district firms (Bagella *et al.*, 1998; Bagella, 2000; Majocchi, 2000; Zucchella, 2002). The impact on international activity of locating in an industrial district has been quantified under the concept of 'district effect'. There is enough empirical evidence to support the influence of district location on internationalization, particularly on its start and intensity. For example, Maccarini *et al.* (2003a, 2003b) find a district effect for export propensity, intensity and geographical spread, both for established and new firms. On the other hand, the evidence is scarce for the Spanish case. An exception is the work of Costa and Viladecans (1999), who analyse the export intensity of firms in 22 manufacturing sectors, finding a positive effect of district membership in ten of those sectors, among them the textile-clothing sector.

3.4 Conclusions

This chapter has discussed the current globalization environment and the internationalization strategy followed by district firms as a response to that environment. The question of departure was whether, given the current challenges of globalization, delocation is becoming a strong strategy of internationalization in industrial districts or, on the contrary, whether internationalization of district firms is dependent on other aspects, such as the industrial organization model, production specialization and, in general, the territorial system of the industry. Although the internationalization of district firms seems to be a natural survival strategy in the globalization environment, the sourcing of foreign suppliers can endanger the continuity of the district system.

The analysis carried out with reference to traditional manufacturing firms, characterized by a small average size and by geographical concentration, has shown that the territorial system inhibits the delocation process, especially in the stages where fragmentation is more difficult or that have a higher value added. As a consequence, although globalization favours processes of delocation for all manufacturing industries, the district location may make the impact of globalization on district firms weaker and more diverse. Another key aspect is the conditioning of district survival by delocation strategies as the rupture of the value chain in the territorial organizational model endangers its evolution.

The influence of territorial systems has also been analysed, especially with respect to foreign markets. Compared to sequential models of internationalization, the network framework is arguably more useful for understanding the process of delocation in industrial districts. From this point of view, internationalization depends more on the set of relationships that the firm develops within the network than on holding any specific advantage.

4 The textile-clothing industry in Europe

Characterization and recent evolution

The worldwide globalization process has brought important challenges to European economies, particularly in traditional manufacturing, of which the textile-clothing industry is a prime example. The complexity of the changes forced upon European countries is, to a great extent, due to their heterogeneity, which increased substantially following the enlargements of May 2004 and January 2007.[1] Moreover, industries themselves are heterogeneous, being composed of several specialized subsectors that differ with respect to product characteristics, market conditions, employment and value added, structural characteristics, degree of specialization and territorially based relationships (Jones, 2002).

The degree of country and industry heterogeneity will become clear when characterizing the current situation of traditional manufacturing in Europe and then focusing on the textile-clothing industry in the Comunidad Valenciana (NUTSII ES52 region in Spain). The choice of industry and geographical area is motivated by four particular features: the historical tendency of the textile-clothing industry to agglomerate in space; the importance of the textile-clothing industry in the regional economy of the Comunidad Valenciana; the level of specialization of the Comunidad Valenciana in the home-textiles subsector; the recent developments in the industry, such as worldwide trade liberalization. These features make it a good case study of future opportunities.

In fact, it has already been asked whether the textile-clothing industry in Europe is evolving or facing extinction (Jones and Hayes, 2004). Other contributions examine the role of firm characteristics, specialization patterns, or geographical agglomeration in the survival possibilities of the European textile-clothing industry (Stenng, 2001; Taplin, 2006). Further research suggests key strategies such as specialization and diversification towards high value added subsectors in order to survive (Buxey, 2005).

The need to restructure the Portuguese textile-clothing industry was analysed by Cristovam (2006). In Portugal, most textile-clothing firms have continued to compete based on low costs both at home and abroad, producing under subcontracting and focusing on the lower-quality end of the market. Given the presence of very low cost producers in countries such as China, this strategy cannot succeed and also perpetuates low wages in the industry. Only a few firms have realized this and started to specialize in higher value added subsectors, focusing on design, quality and service, outsourcing the labour-intensive tasks to lower-cost countries. To most firms, the international outsourcing option is not feasible due to their small size (they had an average size of 11 employees in 2003). Given the very different levels of competitiveness of the firms and their large number, trade associations have found it very difficult to agree on a consensual long-term strategy. Moreover, skill upgrading is a remote option as schooling levels are, on average, very low.

In the case of Spain, the Association of Spanish Textile Firms (ACTE) has recently been promoting the study of the influence of the territory and specialization of production on textile firms, suggesting that a district organization with specialization in high value added subsectors may be a solution. Given the high levels of geographical concentration of the textile industry in Europe, reorganization within the industry may be the most feasible option.

In this context, this chapter analyses the current difficulties of the textile-clothing industry in Europe and points towards an understanding of the implications of specializing on the textile-clothing subsector that is prevalent in the territory where a firm is located. The first part of the chapter studies the main characteristics of the industry, looking at the evolution of demand and specialization patterns. The second part of the chapter analyses the importance of the industry in the EU and, bearing in mind the study that will be presented in the following two chapters, focuses on the case of Spain. Along the same lines, the third part of the chapter examines the presence of the industry in the Comunidad Valenciana from the point of view of the territory and the specialization patterns. Finally, the main conclusions are summarized.

4.1 Characterization of the textile-clothing industry

The textile-clothing sector is one of the major sectors of traditional manufacturing, defined as the set of labour-intensive industries currently in a mature or declining stage in high-income economies.[2] The European Commission defines the textile-clothing sector as

> a diverse and heterogeneous industry which covers a wide variety of products from hi-tech synthetic yarns to wool fabrics, cotton bed linen to industrial filters, or nappies to high fashion. This diversity of end products corresponds to a multitude of industrial processes, enterprises or market structures.
>
> (European Commission, 2009)

The textile-clothing sector is recognized as ‘an important part of European manufacturing industry’, accounting for 4 per cent of manufacturing value added and 7 per cent of manufacturing employment in the EU-15.

A first look at the sector can be taken by distinguishing the demand from the supply side of the market. The demand analysis takes into account the evolution of market demand, while the supply analysis considers production efficiency and competitiveness. Similarly to the rest of traditional manufacturing, the textile-clothing industry is a mature or declining industry in high-income countries, and so domestic demand in these countries is increasingly satisfied by imports (Kilduff and Chi, 2006). On the other hand, the textile-clothing industry has the lowest human capital intensity among European manufacturing industries (Baldwin *et al.*, 2000). With low levels of human capital and technology, product and process innovation is reduced. As a consequence, the potential for market expansion is low (Christoffersen, 2002).

According to the Pavitt (1984) classification, traditional manufacturing is said to be dominated by suppliers in the sense that the suppliers of capital and intermediate goods are the main sources of innovation. Hence, traditional manufacturing has low levels of internalized R&D as it depends on other sectors for innovation (Dalziel, 2007). For traditional manufacturing as a whole, the most common forms of innovation are process innovation and improvements to existing products. For the textile-clothing industry, Tomás and Torrejón (2000) found low levels of investment in R&D activities in territorial systems of SMEs. Moreover, the structure of R&D activities is shared between product development (63 per cent), new materials (27 per cent) and process innovation (17 per cent).

In its characteristics of low demand growth and low technological levels, the textile-clothing industry is representative of the average behaviour of traditional manufacturing.

4.1.1 Evolution of demand

In general, the textile-clothing industry can be characterized with respect to clients, markets and products. First, with respect to clients, to be a mature industry means that clients know the industry's products and have matured in the evaluation they make of them. Second, the two main motivations of demand for final products are replacement-buying and impulse-buying. Both motivations are sensitive to business cycles, particularly impulse-buying. Third, the distinction between materials, intermediate products and final products is blurred, and the relationship between producers and distributors changes rapidly. These phenomena change the configuration of the industry and reinforce the role of industrial associations.[3] Fourth, the textile-clothing industry operates in an open economy where, although quotas were dismantled in January 2005, there are still high tariffs and non-tariff barriers of various kinds. Finally, the growing importance of intangible aspects such as design and branding, without property rights protection, makes the product liable to counterfeiting.

At the same time, the share of household income spent on clothing-footwear is declining (Table 4.1). According to Eurostat,[4] in 1999 EU-15 families spent 6.1 per cent of their income on clothing and footwear (7.4 per cent in Spain), but in 2005 only 5.8 per cent was spent in this way (6.9 per cent in Spain). This change corresponds to a declining share of clothing and footwear expenditure in income of –4.9 per cent in the EU-15 and of –6.8 per cent in Spain, which clearly shows the low income elasticity of mature industries.

Table 4.1 Structure of household expenditure in the EU-15 (%, 1999–2005)

	EU-15			*Spain*		
	1999	*2005*	*Change*	*1999*	*2005*	*Change*
Food and beverages	16.5	15.8	–4.2	21.0	20.3	–3.3
Clothing and footwear	6.1	5.8	–4.9	7.4	6.9	–6.8
Housing and housing-related expenditure	34.6	34.2	–1.2	32.5	35.0	7.7
Health and education	3.9	4.1	5.1	3.9	3.3	–15.4
Transport and communications	15.8	15.5	–1.9	14.5	13.2	–9.0
Recreation and restaurantion	16.0	14.9	–6.9	15.4	15.7	1.9
Others	7.0	9.6	37.1	5.1	5.8	13.7

Source: constructed using data from Eurostat (2009).

Moreover, as has been argued, the evolution of demand and innovation is not uniform across the whole textile-clothing industry. On the contrary, the various subsectors are rather heterogeneous. In order to understand the role of the textile-clothing subsectors, the entire production process has to be described in greater detail and the various subsectors need to be clearly identified.

4.1.2 Definition of subsectors

The EU classification of the various textile-clothing subsectors is summarized in Table 4.2. The textile-clothing industry in a strict sense is represented by codes 17 and 18 of NACE Rev. 1.1.[5] In this way, related activities, such as artificial fibres (code 24.7) and leather and footwear (code 19), will not be included in the subsequent analysis. The full details of the activities included in those codes are listed in Table 4.3. Examining the nature of the subsectors, both 18.1 and 18.3 are very dependent on other industries and use very distinct materials and processes. Moreover, they represent only 2.5 per cent of sector 18's output. As a consequence, subsectors 18.1 and 18.3 will be excluded from subsequent analysis.

Table 4.2 The textile-clothing industry in NACE Rev. 1.1

Generic classification	*Description*	*Codes*
Textiles, leather, footwear and clothing	Textiles	17.1, 17.2, 17.3, 17.6, 17.7
	Leather	19.1, 19.2
	Other textile-clothing industries	17.4, 17.5
	Footwear	19.3
	Clothing and furs	18.1, 18.2, 18.3

Source: constructed using data from Eurostat (2009).

Table 4.3 Activities included in NACE Rev. 1.1 codes 17 and 18

Textile-Clothing Industry	
17	**Textiles**
17.1	**Preparation and spinning of textile fibres** 17.11 – Preparation and spinning of cotton-type fibres 17.12 – Preparation and spinning of woolen-type fibres 17.13 – Preparation and spinning of worsted-type fibres 17.14 – Preparation and spinning of flax-type fibres 17.15 – Throwing and preparation of silk, including from noils, and throwing and texturing of synthetic or artificial filament yarns 17.16 – Manufacture of sewing threads 17.17 – Preparation and spinning of other textile fibres
17.2	**Weaving of textiles** 17.21 – Cotton-type weaving 17.22 – Woolen-type weaving 17.23 – Worsted-type weaving 17.24 – Silk-type weaving 17.25 – Other textile weaving
17.3	**Finishing of textiles** This class includes: • bleaching, dyeing and printing (including thermoprinting) of not-self-produced textile fibres, yarns, fabrics and textile articles, including wearing apparel • dressing, drying, steaming, shrinking, mending, sanforizing, mercerizing of not-self-produced textiles and textile articles, including wearing apparel • finishing of leather wearing apparel • bleaching of jeans • pleating and similar work on textiles

continued

Table 4.3 Continued

Textile-Clothing Industry	

17.4	**Manufacture of made-up textile articles, except apparel**
	This class includes:
	• manufacture of made-up articles of any textile material, including of knitted or crocheted fabrics:
	  • blankets, including travelling rugs
	  • bed, table, toilet or kitchen linen
	  • quilts, eiderdowns, cushions, pouffes, pillows, sleeping bags, etc.
	• manufacture of made-up furnishing articles:
	  • curtains, valances, blinds, bedspreads, furniture or machine covers, etc.
	  • tarpaulins, tents, camping goods, sails, sun blinds, loose covers for cars, etc.
	  • flags, banners, pennants, etc.
	  • dustcloths, dishcloths and similar articles, life jackets, parachutes, etc.
	• manufacture of the textile part of electric blankets

17.5	**Manufacture of other textiles** 17.51 – Manufacture of carpets and rugs 17.52 – Manufacture of cordage, rope, twine and netting 17.53 – Manufacture of non-wovens and articles made from non-wovens, except apparel 17.54 – Manufacture of other textiles

17.6	**Manufacture of knitted and crocheted fabrics**
	This class includes:
	• manufacture and processing in the same unit of knitted or crocheted fabrics:
	  • pile and terry fabrics
	  • net and window furnishing type fabrics knitted on Raschel or similar machines
	  • other knitted or crocheted fabrics

17.7	**Manufacture of knitted and crocheted articles** 17.71 – Manufacture of knitted and crocheted hosiery 17.72 – Manufacture of knitted and crocheted pullovers, cardigans and similar articles
18	**Clothing and furs**
18.1	**Manufacture of leather clothes** This class includes: • manufacture of wearing apparel made of leather or composition leather
18.2	**Manufacture of wearing apparel and accessories** 18.21 – Manufacture of workwear 18.22 – Manufacture of other outerwear 18.23 – Manufacture of underwear 18.24 – Manufacture of other wearing apparel and accessories

18.3	**Dressing and dyeing of fur; manufacture of articles of fur**
	This class includes:
	• dressing and dyeing of furskins and hides with the hair on: scraping, currying, tanning, bleaching, shearing and plucking and dyeing of furskins
	• manufacture of articles made of furskins:
	  • fur wearing apparel and clothing accessories
	  • assemblies of furskins such as ‘dropped’ furskins, plates, mats, strips, etc.
	  • diverse articles of furskins: rugs, unstuffed pouffes, industrial polishing cloths
	• manufacture of artificial fur and articles thereof

Source: Eurostat (2009).

4.1.3 Production process

The production process of the textile-clothing industry includes all the tasks from the first processing of natural and artificial textile fibres to their transformation into cloth which is the basis for a variety of final and intermediate products (Toyne *et al.*, 1984). In general, intermediate textile products that are used as inputs into other products are described in subsectors 17.1, 17.2 and 17.3. Final textile products (home and technical textiles) and wearing apparel (clothing) are included in subsectors 17.4, 17.5, 17.6, 17.7 and 18.2. In 2004–07, clothing output represented almost 30 per cent of total textile-clothing production in the EU-27 and almost half of Spain's production in that industry, with a growing share (Table 4.4). The second and third places were taken by weaving (17.2) and technical textiles (17.5), although weaving is a declining subsector, while the production of technical textiles has grown by 5.3 per cent in the EU-27 and by 12.1 per cent in Spain.

The textile-clothing production process can be grouped into three main cycles as represented in Figure 4.1: production of inputs, transformation of inputs and production of outputs. In recent years the structure of textile-clothing production in Europe has moved up the value chain, away from the initial stages that rely on low labour costs and into later stages that are more intensive in technology and design (Table 4.4).

The input cycle aggregates the activities that take place prior to the manufacture of textile products. The main input of the textile-clothing industry is textile fibres, which receive contributions also from the chemical and machinery industries. The synthetic fibres, obtained from derivatives of petroleum through chemical processes, represent a high share of the total consumption of fibres in the textile industry and their use has been growing in importance, with strategic consequences for the industry. Whereas the degree of industrial concentration (at the level of market structure) is low among the producers of natural fibres, the producers of synthetic fibres follow an oligopolistic, or in some cases monopolistic, market structure, facilitated by capital and technology intensity. These producers

Table 4.4 Structure of textile-clothing production (%, 2004–07)

Codes	*Description*	*EU-27*		*Spain*	
		Share	*Change*	*Share*	*Change*
17.1	Preparation and spinning of textile fibres	9.1	−16.3	8.2	−34.7
17.2	Weaving of textiles	16.9	−5.2	13.2	−8.9
17.3	Finishing of textiles	7.1	14.8	5.6	−7.4
17.4	Manufacture of made-up textile articles, except apparel	9.8	6.8	9.4	13.8
17.5	Manufacture of other textiles	19.0	5.3	9.7	12.1
17.7	Manufacture of knitted and crocheted articles	8.5	−0.3	5.4	0.2
18.2	Manufacture of wearing apparel and accessories	29.6	0.0	48.6	6.3

Source: constructed using data from Eurostat (2009).

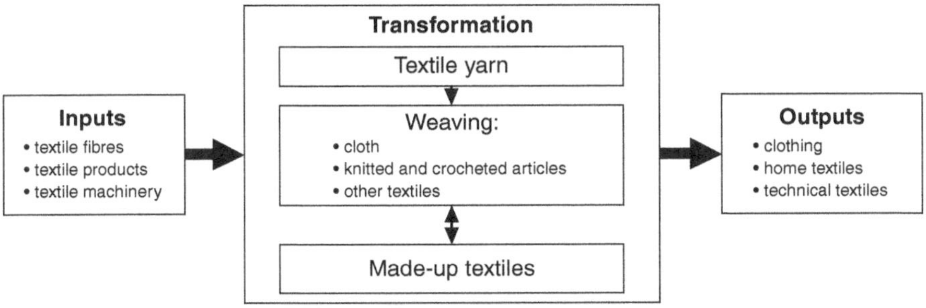

Figure 4.1 The textile-clothing production process.

have high economies of scale, resulting in the concentration of production in a handful of large chemical-products MNEs which possess high bargaining power (Golf *et al.*, 2005).

The specialized textile-clothing machinery also supervises the quality control and design processes, and in this way contributes towards a great deal of incremental process innovation in the textile-clothing industry. The five largest exporters of specialized textile-clothing machinery in 2004–07 were Germany, Japan, Italy, Switzerland and China (Comtrade, 2009). Other countries that are important textile-clothing producers, such as Spain and Portugal, are in this way dependent to some extent on the technological developments carried out by machinery exporters. In general, the textile input cycle is dependent on suppliers, which sometimes hold a significant degree of market power. The degree of dependency is greatest in the most 'traditional' segments – those with a high product standardization, which favours the concentration of market structure and geographical specialization.

In the intermediate segment of the textile manufacturing process (transformation), textile articles are manufactured by means of weaving, knitting and finishing. Technological progress in the development of synthetic fibres and manufacturing processes has been replacing weaving with other techniques for obtaining non-weaved and knitted articles that allow resistance to wear and tear, prevent the loss of body heat or protect from sunlight, and provide applications to interior decoration, agriculture or industry. The finishing process of eliminating defects from cloth and changing its overall look occurs at different stages of textile manufacturing (Detrell, 1998).

Finally, the textile manufacturing cycle closes with the manufacturing of apparel, which transforms cloth into an output. The variety of products obtained can be classified into three broad categories: clothing, home textiles and technical textiles. The production of technical textiles has grown in recent years, placing this subsector as the second largest in the EU-27 (third largest in Spain), with a share of 19 per cent (9.7 per cent) of textile-clothing production (Table 4.4). There is an increasing variety of uses for technical textiles, which are described in Box 4.1.

Box 4.1 Uses for technical textiles

1 Textiles for civil engineering: hydraulics and drainage systems.
2 Textiles for agriculture and fishing: greenhouses and netting.
3 Textiles for construction: isolators and wallpaper.
4 Textiles for medical and hygenic uses.
5 Textiles for environmental protection: residue containers.
6 Textiles for vehicles, transport and packaging: tyres, bags and containers.
7 Textiles for protection: reflectors and life jackets.
8 Textiles for sports and outdoors: sleeping bags and sails.
9 Textiles for industrial uses: conveyer belts and hose pipes.

Source: Adapted from Detrell (1998).

In sum, the textile-clothing industry is highly dependent on other sectors, particularly for innovation; the demand for textile-clothing products has not increased on par with income; and the industry is made up of various segments with very distinct characteristics.

4.2 The recent evolution of the textile-clothing industry in Europe

4.2.1 Evolution of the textile-clothing industry in the EU-27

In 2004, industrial employment[6] represented around 28 per cent of total employment in the EU-25, with 45.16 million workers (Eurostat, 2009). In turn, in 2001–05, textile-clothing employment represented around 40 per cent of industrial employment in the EU-27, with over two million workers (Eurostat, 2009). The high share of textile-clothing in industrial employment is understandable if it is taken into account that it is the most labour-intensive industrial branch in the EU.

The textile-clothing industry is a source of large imbalances in world trade, with Europe and America generally recording deficits and Asia having a growing surplus. In 2000, the two largest textile markets (Europe and America) each manufactured 29 per cent of the total world production, while the third largest market (Asia) manufactured 39 per cent (Stengg, 2001). Similarly for clothing, Asia produced 45 per cent of the total world production and Europe and America produced 26 per cent each (Figure 4.2).

In 2004–07, the export markets were dominated by a handful of countries, and the prominent role of China in a wide range of subsectors was noticeable (Table 4.5). Italy had the second widest presence across a range of subsectors, including inputs, intermediates and outputs. Other EU countries such as Germany, France and Belgium have specialized in particular product niches in association with other industries that are strong domestically. Germany, for example, was the largest world exporter of non-wovens and special yarns, and

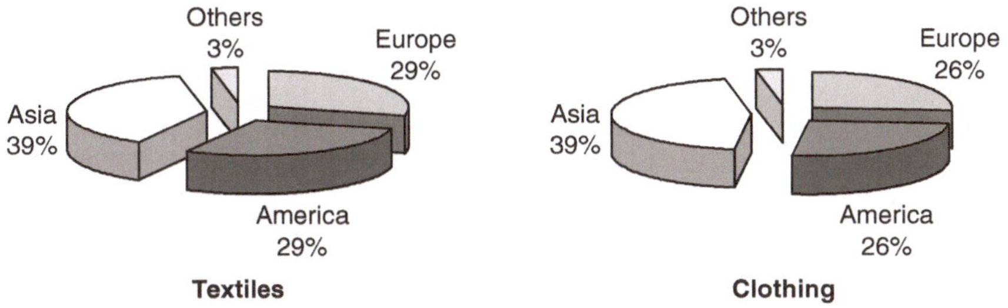

Figure 4.2 Shares of world production in the textile-clothing industry (%, 2000) (source: adapted from Stengg (2001)).

the second largest world exporter of man-made filaments and staple fibres. Germany's strong presence in the technical textiles subsectors is sustainable through close collaboration with the chemical industry and a heavy investment in R&D. In general, emerging markets such as China tend to dominate world markets in standard labour-intensive low-technology products, while some EU countries such as Germany have succeeded in maintaining a strong presence in innovative technology-intensive products.

The textile-clothing industry in the EU-27 had a total production of over €117 million in 2007 (Table 4.6) and employed almost 2.2 million people in 2006 (Table 4.7). Among the EU-27 member states there are important differences regarding the share of the industry in their economies in terms of production and employment, as well as the subsector structure of production.

Italy alone accounts for over 35 per cent of textile-clothing production in the EU-27, with Germany and Spain accounting for over 9 per cent each. It is interesting to note that textile-clothing production in the EU has halved since 2001, but this knock-down effect has taken place mostly in the clothing sector. Spain was the third largest textile producer in the EU-27 in 2007, but the second largest in clothing, with a share of over 15 per cent of the EU-27's clothing production. In other EU countries such as Germany, France and Belgium, which are specialized in niche products with a high technology and design content, textile production dwarfs clothing production. The opposite extreme can be observed in Romania, where clothing production is more than three times larger than textile production, due to the delocation of labour-intensive stages of production by other EU countries (Smith *et al.*, 2005). Unlike the two largest textile-clothing producers in the EU – Italy and Germany – Spain is not among the world's largest exporters. In reality, Spain's production is biased towards domestic consumption, whereas Italy and Germany produce mainly to export in particular market niches.

Italy, Romania and Poland are the leading countries in terms of employment, with around 22 per cent, 15 per cent and 11 per cent of the EU-27 total respectively. Comparing employment and production data, it becomes apparent which

Table 4.5 Textile-clothing world export share (2004–07)

Codes	*Description*	*Country*	*Share (%)*
H2-50	Silk	China	42.10
		Italy	14.27
		India	11.94
H2-51	Wool	Italy	20.89
		Australia	14.38
		China	14.20
H2-52	Cotton	China	17.94
		USA	13.94
		Hong Kong	11.33
H2-53	Other vegetable textile fibres	China	17.88
		Italy	12.41
		France	11.17
H2-54	Man-made filaments	China	18.81
		Germany	8.89
		Korea	8.89
H2-55	Man-made staple fibres	China	19.22
		Germany	9.39
		USA	7.26
H2-56	Non-wovens and special yarns	Germany	13.56
		USA	11.56
		Italy	9.76
H2-57	Carpets and other textile floor coverings	Belgium	21.75
		India	9.21
		Netherlands	9.02
H2-58	Special woven fabrics	China	28.10
		Hong Kong	8.08
		USA	6.95
H2-59	Impregnated, coated, covered or laminated textile fabrics	Germany	13.04
		China	12.76
		USA	9.58
H2-60	Knitted or crocheted fabrics	China	22.19
		Hong Kong	15.67
		Korea	15.01
H2-61	Articles of apparel and clothing accessories, knitted or crocheted	China	32.15
		Hong Kong	10.59
		Italy	5.68
H2-62	Articles of apparel and clothing accessories, not knitted or crocheted	China	27.78
		Hong Kong	8.84
		Italy	8.32
H2-63	Other made-up textile articles	China	32.44
		Pakistan	8.74
		India	6.55

Source: constructed using data from Comtrade (2009).

EU-27 countries have specialized in relatively labour-intensive subsectors. Romania, Poland, Portugal and the Czech Republic all have a higher share of EU-27 employment than of EU-27 production, and in all cases the clothing subsector dominates the textile subsector. These countries are clearly the most sensitive to the worldwide liberalization of textile-clothing trade that opened up the EU market to low-labour-cost emerging markets. Sensitiveness is even higher if taking into account the high levels of regional specialization in regions of Belgium, Bulgaria, Hungary, Italy, Poland, Portugal and Romania as represented by NUTSII regions with textile-clothing specialization coefficients[7] that are over twice the EU-27 average (Table 4.8). Bulgaria and Portugal include the two EU-27 regions (Severen tsentralen and Norte, respectively) with the highest textile-clothing specialization coefficient – more than six times the EU-27 average.

Table 4.6 Textile-clothing production in the EU (2007)

	Textiles (NACE 17)		*Clothing (NACE 18)*		*Total production*	
	Million €	*Share (%)*	*Million €*	*Share (%)*	*Million €*	*Share (%)*
Italy	28,456	34.06	12,748	37.85	41,204	35.15
Germany	9,789	11.72	1,066	3.16	10,855	9.26
Spain	5,537	6.63	5,245	15.57	10,783	9.20
United Kingdom	6,426	7.69	2,003	5.95	8,430	7.19
France	6,559	7.85	523	1.55	7,082	6.04
Romania	1,029	1.23	3,622	10.75	4,651	3.97
Portugal	2,553	3.06	1,919	5.70	4,472	3.82
Belgium	3,798	4.55	133	0.39	3,931	3.35
Poland	1,750	2.09	1,071	3.18	2,820	2.41
EU-27	83,540	100	33,681	100	117,221	100

Source: constructed using data from Eurostat (2009).

Table 4.7 Textile-clothing employment in the EU (2006)

	Textiles (NACE 17)		*Clothing (NACE 18)*		*Total production*	
	Million €	*Share (%)*	*Million €*	*Share (%)*	*Million €*	*Share (%)*
Italy	234	23.56	238	19.88	471	21.51
Romania	71	7.15	257	21.47	327	14.95
Poland	81	8.16	152	12.70	234	10.66
Portugal	76	7.65	110	9.19	186	8.51
Spain	82	8.26	93	7.77	176	8.02
France	79	7.96	58	4.85	137	6.26
Germany	83	8.36	41	3.43	124	5.65
United Kingdom	76	7.65	34	2.84	109	4.98
Czech Republic	46	4.63	33	2.76	80	3.64
EU-27	993	100	1,197	100	2,190	100

Source: constructed using data from Eurostat (2009).

Traditionally, the textile-clothing industry has benefited from trade policy protection in high-income countries. The most important mechanism set up at the worldwide level has been, until 2005, the Multifiber Agreement, called the Agreement on Textiles and Clothing (ATC) since the WTO replaced the GATT in 1995. This system of quotas prevented, to a large extent, imports from Asian countries, which could not be effectively prevented by tariffs due to the very low labour costs at the origin, as proved by the noticeable expansion of Asian exports since 2005, when the ATC was finally dismantled. Due to the political sensitiveness of trade liberalization in the textile-clothing sector, the ATC effectively established several stages of liberalization from 1995 to 2005 by means of the progressive reduction of quotas imposed by the EU and the United States on the exports of Argentina, China (including Hong Kong and Macao), Korea, the Philippines, India, Indonesia, Malaysia, Pakistan, Peru, Singapore and Thailand. In 1995, 16 per cent of textile-clothing trade was liberalized. In 1998, another 17 per cent was liberalized, followed by 18 per cent more in 2002. The remaining 49 per cent was finally liberalized in 2005.

Table 4.8 Coefficients of specialization, number of firms and their employees in the EU-27 NUTSII regions most specialized in textile-clothing (2002–05)

			CS	*Number of firms*	*Number of employees*
Belgium	be25	Prov. West-Vlaanderen	2.822	707	22,347
Bulgaria	bg31	Severozapaden	2.207	603	164
	bg32	Severen tsentralen	6.147	404	2,758
	bg33	Severoiztochen	3.354	282	654
	bg41	Yugozapaden	3.899	1,282	37,797
	bg42	Yuzhen tsentralen	5.383	1,374	27,150
Hungary	hu22	Nyugat-Dunantul	2.646	237	7,078
	hu32	Eszak-Alfold	2.063	307	4,615
Italy	itc4	Lombardia	2.504	3,800	116,577
	itd3	Veneto	2.698	1,595	56,338
	ite1	Toscana	3.278	2,832	35,060
	ite2	Umbria	2.594	304	5,181
	ite3	Marche	2.162	516	11,591
	itf1	Abruzzo	2.591	493	10,784
Poland	pl11	Lodzkie	3.821	181	22,824
Portugal	pt11	Norte	6.438	2,623	28,172
Romania	ro11	Nord-Vest	3.214	1,432	50,682
	ro12	Centru	3.933	1,172	53,166
	ro21	Nord-Est	2.842	1,533	72,695
	ro22	Sud-Est	2.958	1,011	49,029
	ro31	Sud – Muntenia	2.317	830	40,903
	ro32	Bucuresti – Ilfov	2.939	3,270	92,720
	ro42	Vest	3.176	830	38,435

Source: constructed using data from Eurostat (2009) and AMADEUS (2008).

The EU's lack of competitiveness in the textile-clothing sector has become clear as liberalization has progressed, with its deficit increasing from –€30 billion in 2004 to –€49 billion in 2007 (Figure 4.3). The EU-27 has a small surplus in textile products (around €3 billion), but a very large and increasing deficit in clothing, which caused the increase in the overall deficit. The same trend has been followed by the new member countries in Central and Eastern Europe (CEECs), which are a major destination for the EU-15's textile products. This pattern of trade has its roots in the Europe Agreements, a set of bilateral free trade agreements signed between the EU-15 and each of the CEECs during the 1990s (Spies and Marques, 2009). The Europe Agreements exempted some 'sensitive' sectors from tariffs, among them textile-clothing, if the product had a minimum content of EU input. Through outsourcing of labour-intensive stages of production to the CEECs, many EU-15 textile-clothing firms ensured their textile inputs would be transformed and re-exported as finished goods free of tariffs.

Table 4.9 shows the main trade partners of the EU-15 in 2006 for both exports and imports of textile-clothing.[8] Together with some high-income countries – such as Switzerland, the United States and Japan – Bulgaria and Romania are major trade partners, as well as Mediterranean countries such as Morocco. However, exports to high-income countries are mostly comprised of clothing, whereas exports to middle-income countries are mostly comprised of textile

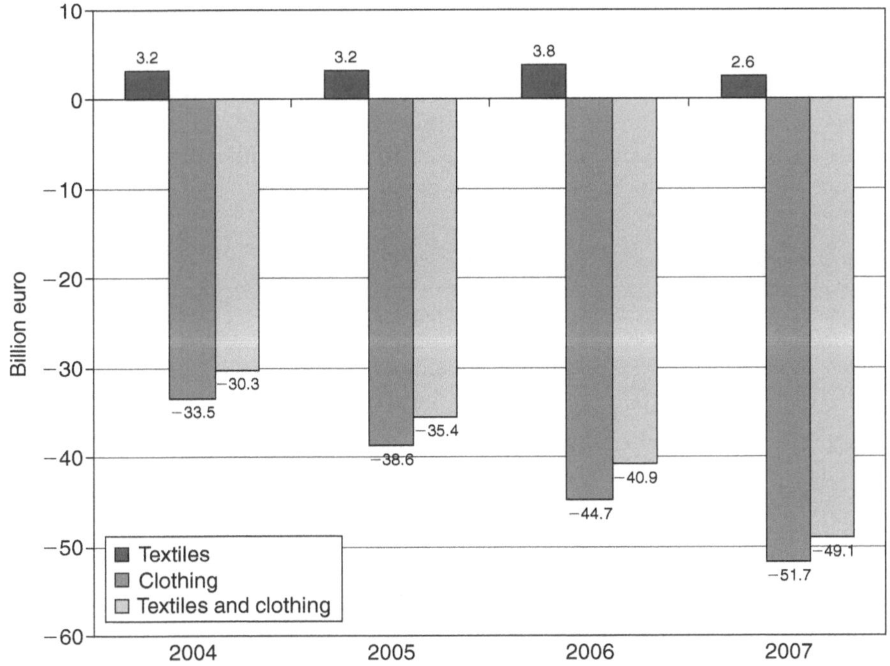

Figure 4.3 EU-27 textile-clothing trade balance (€ billions, 2004–07) (source: constructed using data from Comtrade (2009)).

inputs. On the other hand, most imports originate from middle-income countries in Asia and the Mediterranean, particularly from China, and from Bulgaria and Romania. Imports are also two times more concentrated than exports, with the 12 main partners taking 34.2 per cent of imports and 18.71 per cent of exports. This is due, to a large extent, to the very high shares of China (12.61 per cent), Turkey (5.93 per cent) and India (3.2 per cent) in EU-15 imports, which surpass the US share in EU-15 exports (3.14 per cent).

The two parallel processes of enlargement to the East and liberalization of the worldwide textile-clothing trade have been putting pressure on the EU's textile-clothing industry. The impact of liberalization was felt on trade patterns through the growth of exports from China and other important suppliers, reinforcing the EU's trade deficit in this industry: in 2004–06, China's share in EU-15 imports increased from 8.7 per cent in 2004 to 12.1 per cent in 2005 and 12.6 per cent in 2006 (Comext, 2009). Liberalization also impacted on industrial structure through the exit of the weakest firms and ensuing increase in concentration. As a consequence, the EU faces production and employment losses in textile-clothing, which might not be sufficiently compensated by gains in downstream industries (in retail, for example) and might be extended by a rigid labour market that makes sectoral reallocation of workers more difficult.

Observing the rapid increase in imports from China with a negative impact on EU textile-clothing firms, in 2005 the EU decided to reinstate some restrictions under the form of temporary decreasing quotas and 'prohibitive' tariffs, which dampened the growth of China's import share in 2006. However, Table 4.10 shows that the decline in the EU's textile-clothing industry – as represented by higher trade deficits, exit of firms, loss of production and jobs – had existed before the liberalization of worldwide trade in 2005 and the enlargement of 2004. Therefore, the data point to a reduced role of trade liberalization and integration on the decline of the EU's textile-clothing industry.

Table 4.9 EU-15 main textile-clothing trade partners (%, 2006)

Exports	%	*Imports*	%
United States	3.14	China	12.61
Switzerland	2.57	Turkey	5.93
Poland	2.06	India	3.20
Russia	1.95	Bangladesh	2.61
Romania	1.78	Romania	2.07
Turkey	1.39	Tunisia	1.48
Japan	1.27	Hong Kong	1.38
Tunisia	1.19	Morocco	1.35
Hong Kong	1.10	Pakistan	1.22
Morocco	1.06	Indonesia	0.99
Norway	0.68	Bulgaria	0.77
Ukraine	0.51	Vietnam	0.60
Top 12	18.71	Top 12	34.20

Source: constructed using data from Comext (2009).

Table 4.10 The EU's textile clothing sector (2001–04)

	2001	*2004*	*Variation (%)*
Export/import ratio (%)	54.7	52.0	–5.0
Production (€ million)	203,283	180,500	–11.2
Employment (thousands)	2,669	2,263	–15.2
Firms (thousands)	230	165	–27.0

Source: adapted from Stengg (2001), Sura (2004) and Marchi (2005).

As a consequence, the root of that decline also has to be looked for in structural causes rather than only in trade liberalization and integration. Given the growing share of world trade of the textile-clothing industry and the high labour-intensity of the sector that has led firms to delocalize to low-labour-cost countries, some restructuring is called for. However, that restructuring must be adapted to the characteristics of the various subsectors within the industry (textiles are more capital intensive, whereas clothing is more labour intensive) and to the conditions present in each territory (the levels of activity and agglomeration differ across the EU).

4.2.2 The textile-clothing industry in Spain: similarity to the EU-27 and differentiated characteristics

It can be argued that the textile-clothing industry in Spain shares many characteristics with that of the EU-27 as a whole. It is an important industry within traditional manufacturing, very concentrated geographically but based on SMEs, with a deficit in world trade and highly dependent on European markets. Above all, it has been feeling the same symptoms of decline since the start of the twenty-first century.

Spanish firms tend to specialize in only a few textile or clothing products, so there are few vertically integrated firms, and a large proportion of firms are SMEs. Effectively, in 2001–06, microfirms and small firms accounted for respectively 56.5 per cent and 37.2 per cent of textile-clothing firms in Spain, while medium and large firms took up only a residual share of employment in the sector (Figure 4.4).[9]

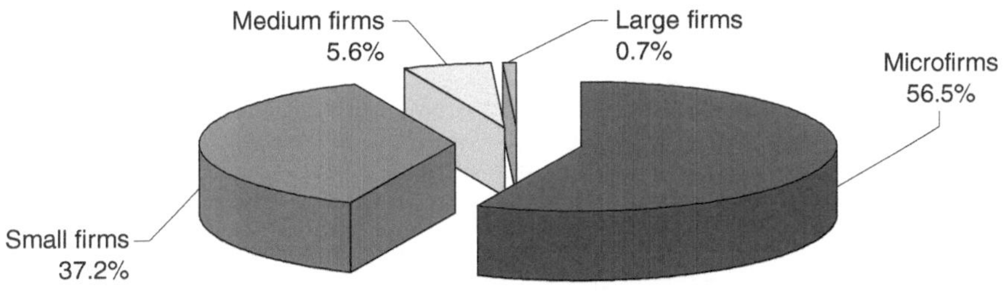

Figure 4.4 Size distribution of textile-clothing firms by number of workers (2001–06) (source: constructed using data from AMADEUS (2008)).

As a consequence of the very small size of most firms and their high specialization level, Spanish textile-clothing firms tend to agglomerate geographically, with a high degree of production interdependence. This tendency has been developed over time and is engrained historically at the European level as well. In 2005, manufacturing represented 13.7 per cent of total employment in Spain, with the textile-clothing industry representing 7.2 per cent of manufacturing employment (Table 4.11). However, the two adjacent regions of Cataluña and Comunidad Valenciana represented 52.7 per cent of total textile-clothing employment in Spain. Together with Castilla-la-Mancha and Galicia, Cataluña and Comunidad Valenciana are the four Spanish regions with a specialization level between 55 per cent and 107 per cent above the Spanish average, as measured by the specialization coefficients relative to the Spanish average. The four regions of Cataluña, Galicia, Castilla-la-Mancha and Comunidad Valenciana had 3,942, 644, 675 and 1,919 textile-clothing firms, respectively. Due to a high density of economic activity, these Spanish regions concentrate some of the largest numbers of firms and employees at the EU-27 NUTSII level, and even though they have a diversified economic structure that does not allow specialization levels to be as high as in other EU-27 regions, they are still more specialized in the textile-clothing sector than the EU-27 average.

In Spain it is possible to distinguish three different types of regional situations with respect to the territorial distribution of the textile-clothing industry (Figure 4.5). First, in white, regions where textile-clothing is complementary or residual. Second, in grey (Castilla-la-Mancha and Galicia), regions that are 70 per cent

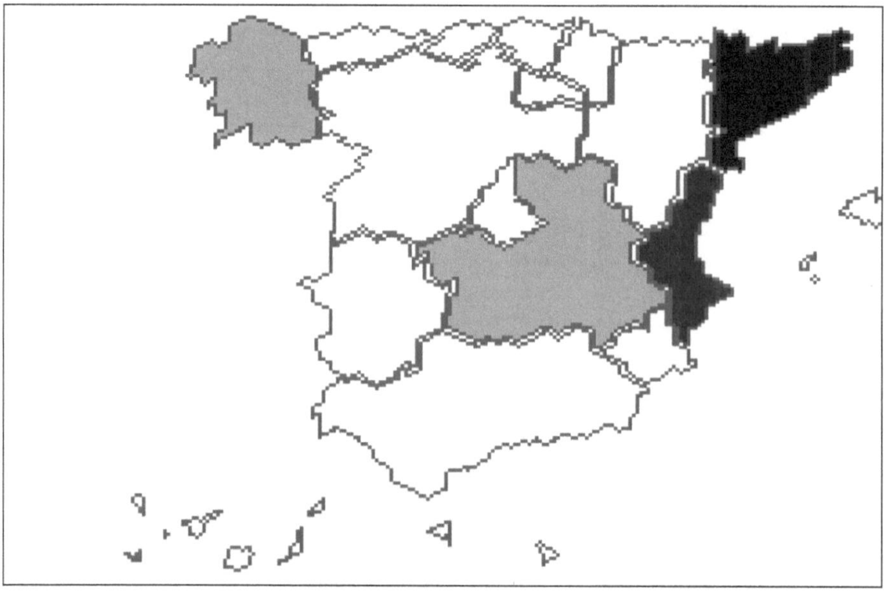

Figure 4.5 Location of the main textile-clothing regional agglomerations in Spain (source: constructed using data from INE (2004)).

Table 4.11 Textile-clothing specialization coefficients of Spanish NUTSII regions (2005)

NUTSII region	*Total employment*	*Textile-clothing employment*	*Share of textile-clothing in manufacturing employment (%)*	*Specialization coefficient*
Cataluña	3,291,100	67,143	10.8	2.07
Galicia	1,130,100	19,773	12.3	1.78
Castilla-la-Mancha	763,700	12,798	10.9	1.70
Comunidad Valenciana	2,053,100	31,319	9.5	1.55
Aragón	568,500	4,583	4.3	0.82
Rioja (La)	140,800	1,100	3.5	0.79
Castilla-León	1,022,700	6,706	4.6	0.67
Murcia (Región de)	569,700	3,296	4.3	0.59
Navarra	274,400	1,600	2.2	0.59
Madrid (Comunidad de)	2,858,800	15,654	5.8	0.56
Andalucía	2,959,600	14,284	5.7	0.49
Extremadura	385,800	1,794	6.0	0.47
Asturias (Principado de)	405,200	1,433	2.6	0.36
Cantabria	239,200	805	2.3	0.34
País Vasco	958,200	2,737	1.2	0.29
Balears (Illes)	468,100	1,024	3.7	0.22
Total	18,973,200	186,837	7.2	1

Source: constructed using data from Eurostat (2009).

more specialized in the industry than the Spanish average but do not each employ more than 20,000 workers in the industry. Finally, in black (Cataluña and Comunidad Valenciana), regions that are also more specialized in textile-clothing than the Spanish average but employ a higher number of workers (over 67,000 and over 31,000 respectively). Given the large numbers involved, both at the relative and absolute levels, the textile-clothing industry has an important impact on these regions.

When considering NUTSIII,[10] the disparities in specialization levels become even higher (Table 4.12). The provinces (NUTSIII) of Barcelona, Valencia and Alicante represent 46 per cent of total textile-clothing employment in the regions (NUTSII) that are highly specialized in textiles-clothing (Castilla-La Mancha, Galicia, Cataluña and Comunidad Valenciana). On the other hand, 11 out of the 16 provinces in those four regions are more specialized in textiles-clothing than the Spanish average. Those provinces that have specialization levels of more than twice the regional average are Toledo, Barcelona, Ciudad Real, Coruña and Albacete. Figure 4.6 represents the five provinces with specialization coefficients higher than 2.0 in black (Toledo, Barcelona, Ciudad Real, Coruña and Albacete) and those with specialization coefficients between 1.0 and 2.0 in grey (Alicante, Girona, Valencia, Ourense, Lleida and Castellón).

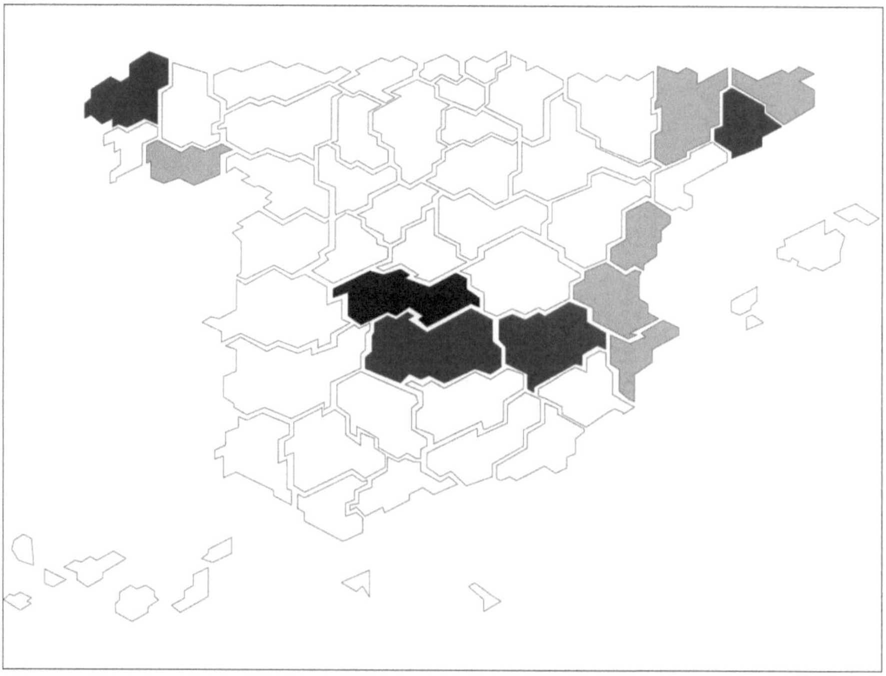

Figure 4.6 Location of the main textile-clothing provincial agglomerations in Spain (source: constructed using data from INE (2004)).

Table 4.12 Textile-clothing specialization coefficients of selected Spanish NUTSIII provinces (2004)

NUTSIII Province	*NUTSII region*	*Total employment*	*Manufacturing employment*	*Textile-clothing employment*	*Specialization coefficient*
Toledo	Castilla-la-Mancha	208,410	51,765	10,758	3.046
Barcelona	Cataluña	2,139,905	554,269	85,513	2.358
Ciudad Real	Castilla-la-Mancha	163,822	27,379	6,269	2.258
Coruña	Galicia	424,910	71,370	16,154	2.243
Albacete	Castilla-la-Mancha	133,204	25,887	4,811	2.131
Alicante	Comunidad Valenciana	585,561	144,450	18,814	1.896
Girona	Cataluña	256,038	55,323	7,997	1.843
Valencia	Comunidad Valenciana	916,303	197,377	22,729	1.463
Ourense	Galicia	115,653	16,555	2,755	1.405
Lleida	Cataluña	158,141	27,305	3,010	1.123
Castellón	Comunidad Valenciana	213,161	61,791	3,965	1.097
Pontevedra	Galicia	359,588	75,826	5,938	0.974
Tarragona	Cataluña	261,042	51,263	3,206	0.724
Cuenca	Castilla-la-Mancha	70,780	9,227	844	0.703
Lugo	Galicia	135,027	15,470	759	0.331
Guadalajara	Castilla-la-Mancha	70,609	11,774	394	0.329
Total		6,212,154	1,397,031	193,916	

Source: constructed using data from INE (2004).

Similarly to the EU-27 as a whole, Spain's textile-clothing exports have grown less than its imports, particularly from 2005, following the worldwide liberalization of the textile-clothing trade (Figure 4.7). As a consequence, Spain faces increasing trade deficits in textile-clothing, with exports representing only 58 per cent of imports in 2007.

Also similarly to other EU-27 countries, Spain's main export markets are neighbouring countries, with textiles behaving differently from clothing (Table 4.13). The EU-27 was Spain's main market and supplier of textiles in 2006. However, imports of clothing were dominated by China (18.7 per cent against Italy's 11 per cent share in the Spanish market). Within the EU-27, Spain's main markets were France, Italy and Portugal, with France and Portugal being more important markets for clothing, and Italy slightly more so for textiles. Outside the EU-27, the main markets were Morocco, Mexico and Turkey, with Morocco being the sole most important market for Spanish textiles (15 per cent against France's 12.4 per cent). As Morocco is also Spain's second largest supplier of clothing, it seems that the textile-clothing trade between Spain and Morocco may be of the vertical type motivated by delocation, whereby Spanish firms export textiles as inputs to Morocco, which exports the transformed clothing output back to Spain. In this way, Spanish firms benefit from closeness to contractors and the low labour costs they can offer. On the other hand, Spain's reliance on Italy and Germany as suppliers of textiles is linked to more specialized niche products, where Spanish firms lie downstream, whether selling in the domestic market or exporting.

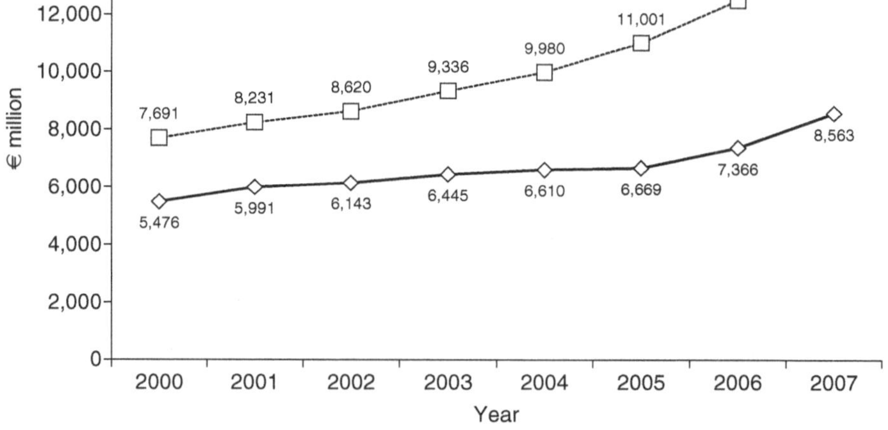

Figure 4.7 Textile-clothing external trade for Spain (€ millions, 2000–07) (source: constructed using data from Comext (2009)).

Table 4.13 Spain's main trade partners in the textile-clothing trade (%, 2006)

	Textile		*Clothing*	
	Exports	*Imports*	*Exports*	*Imports*
EU-27	65.0 France: 12.4 Italy: 10.4 Portugal: 9.7	65.8 Italy: 23.6 Germany: 11.5 France: 10.2	73.4 Portugal: 18.4 France: 16.9 Italy: 9.6	46.1 Italy: 11.0 France: 9.0 Portugal: 8.3
Non EU-27	35.0 Morocco: 15.0 Mexico: 3.1 Turkey: 2.2	34.2 China: 7.8 India: 5.1 Turkey: 5.1	26.6 Mexico: 3.9 Morocco: 3.3 Turkey: 1.8	53.9 China: 18.7 Morocco: 8.8 Turkey: 7.1

Source: constructed using data from Comext (2009).

4.2.3 Current situation of the textile-clothing industry in Spain

The debate on the general loss of competitiveness of the textile-clothing industry in Spain has gravitated around whether its decline is cyclical or structural, and whether it faces restructuring or a crisis (Pla *et al.*, 2007). Most studies treat the difficulties as structural due to the importance of environmental changes affecting the sector. In addition to trade liberalization and European integration, another important change in the textile-clothing sector concerns the simultaneous disappearance of the traditional distribution channels and the concentration on the retail end to the extent of creating, in some cases, a monopsony (Taplin, 2006). In Spain, some notorious examples are retail chain stores such as El Corte Inglés, integrated chains such as Inditex (which produces and sells the Zara and Mango labels), or the clothing retail carried out by hypermarket chains such as Carrefour or Consum.

Moreover, the Spanish textile-clothing industry suffers important strategic and organizational structural weaknesses also present in other traditional manufacturing. First, the under-utilization of interfirm cooperation in spite of the importance of industrial districts. Second, the low investment in marketing and R&D. Third, the excessive atomization. Fourth, the low professionalization of management and lack of definition in the design of organizational structures. Fifth, the reliance on exports and imports as the main means of internationalization, with under-utilization of subcontracting, multilocation and delocation. Finally, the lack of strategic direction in management, which translates into short-termism and reactive decisions that cause under-differentiation of firms, aggravating the crisis as indicated by the evolution of the trade balance, production, employment and number of firms (Pla *et al.*, 2007).

The joint effect of weaknesses and threats led to the loss of production, jobs and firms. In 2001 Spain's textile-clothing exports represented 73 per cent of imports in the same industry, but that value had decreased to 58 per cent in 2007 (Figure 4.8). Note, however, that the single largest annual fall happened between

2004 and 2005, coinciding with the worldwide liberalization of textile-clothing trade, which had two major impacts: the drop in textile exports and the high increase in clothing imports. Production, on the other hand, decreased by around –30 per cent between 2001 and 2007, with important consequences for employment and exit of firms. Employment decreased by –24 per cent between 2001 and 2006. Similarly, between 2000 and 2006 the number of registered firms decreased by –17.19 per cent. The largest exit of firms occurred in 2004 (–5.54 per cent) and 2005 (–6.43 per cent), with strong restructuring in the run-up to the May 2004 EU enlargement and the January 2005 textile-clothing trade liberalization.

When distinguishing the reduction in the number of firms by firm size, the largest drop in 2000–04 occurred in the large (–27.4 per cent) and medium (–27.2 per cent) firm groups (Figure 4.9). This reduction was due not only to the exit of weaker firms, but also to the downsizing that transformed large firms into medium and medium into small.

In conclusion, the textile-clothing sector has been challenged at various levels, resulting in a structural crisis. The decline was particularly severe in 2004–05, and post-2005 data seem to indicate some tendency towards

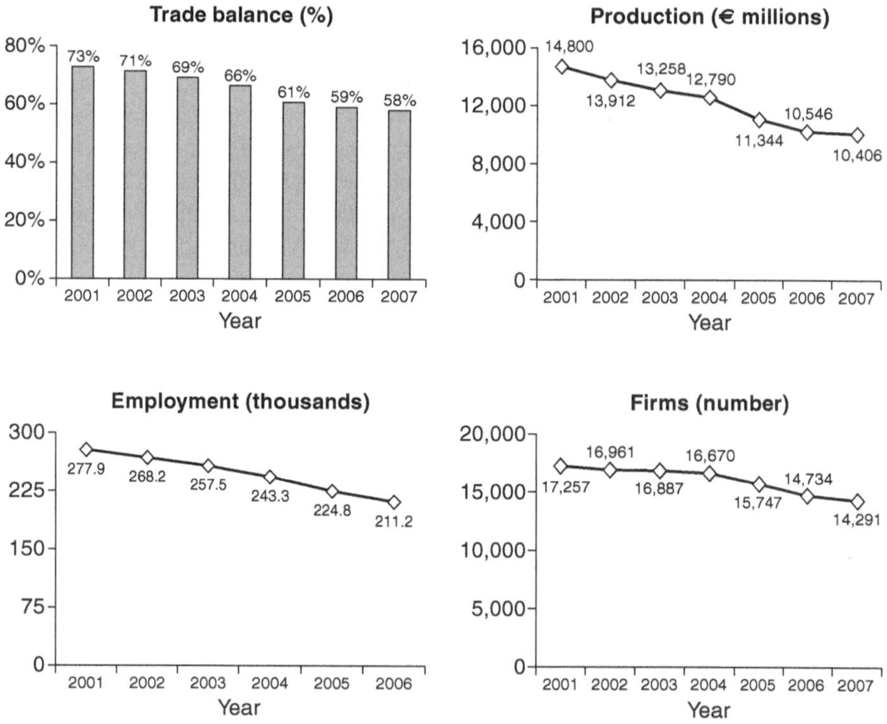

Figure 4.8 Textile-clothing indicators in Spain (sources: Comext (2009), Eurostat (2009)).

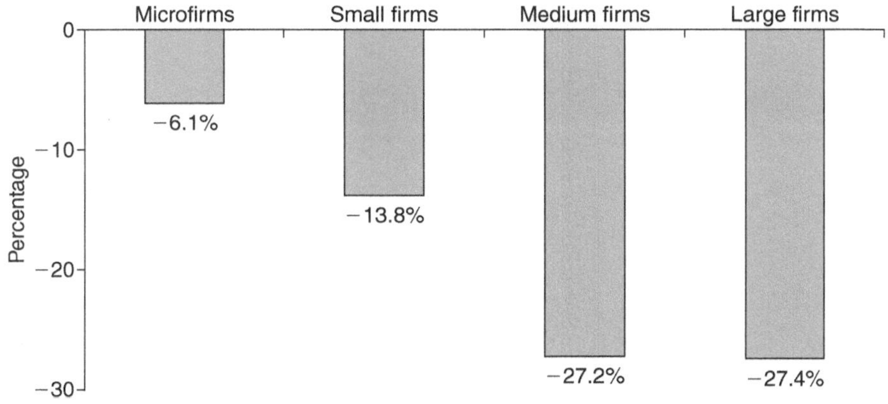

Figure 4.9 Change in the number of textile-clothing firms in Spain (%, 2000–04) (source: constructed using data from INE (2005)).

stabilization at lower levels of activity, perhaps generating a more permanent contraction of the sector. The question is whether the structural crisis spans textile-clothing subsectors and geographical agglomerations equally, or whether some distinctive patterns can be found, whereby the post-liberalization equilibrium has effectively driven out of the market the weaker firms, possibly those with uncompetitive specialization patterns or an unfavourable geographical location and/or organization.

4.3 An example of concentration and territorial specialization in the textile-clothing industry: the Comunidad Valenciana

4.3.1 The economic importance of the textile-clothing industry in the Comunidad Valenciana

The Comunidad Valenciana has a slightly higher share of manufacturing employment than the Spanish average (16.1 per cent against 13.7 per cent in 2005). At the same time, manufacturing value added is 3 per cent higher in the Comunidad Valenciana when compared to the Spanish average, but the region is also strongly specialized in traditional manufacturing, comprising furniture, toys and footwear, as well as textiles-clothing (Boix and Galleto, 2005). Taking the textile-clothing sector as an example, in 2005 it represented 9.5 per cent of manufacturing employment (8.5 per cent of value added) in the Comunidad Valenciana against 7.2 per cent for the Spanish average (Table 4.14). These data make the textile-clothing sector the most important in terms of employment, and the third most important in terms of value added in the Comunidad Valenciana. This strong geographical concentration and specialization of traditional manufacturing, with

Table 4.14 Subsector structure of textile-clothing employment (2000)

Codes	*Description*	*Spain*		*Comunidad Valenciana*		
		Employees	*Share (%)*	*Employees*	*Share (%)*	*Regional share (%)*
17.1, 17.2, 17.3	Textile fibres	37,414	15.0	8,628	20.9	23.0
17.4	Made-up textile articles, except apparel	15,629	6.3	4,480	10.8	28.6
17.5	Other textiles	34,447	13.8	11,621	28.1	33.7
17.7	Knitted and crocheted articles	21,571	8.7	3,077	7.4	14.2
Textile total		109,091	43.8	27,806	67.2	25.3
18.2	Clothing	140,157	56.2	13,570	32.8	9.7
Total		249,248	100	41,376	100	16.6

Source: constructed using data from IVE (2002).

most of the employment in some municipalities concentrated in traditional manufacturing, aggravates the economic and social impact of the decline in those sectors.

The importance of traditional manufacturing in the Comunidad Valenciana can be explained by historical factors related to the region's industrialization between the end of the nineteenth century and the beginning of the twentieth century. The rapid expansion of Valencia's commercial agriculture created high demand for industrial products, both as agricultural inputs and for domestic consumption. The industrial network grew based on family SMEs. In the 1970s the level of industrialization of the Comunidad Valenciana reached those of Cataluña and the Basque Country, helped first by domestic demand and soon by the internationalization of its firms, followed by strong FDI inflows. The historical evolution of industry in the region explains the specialization in traditional manufacturing, based on local SMEs, with strong export intensity and organized in specialized industrial districts (Piqueras, 1999).

4.3.2 Characterization of the textile-clothing industry in the Comunidad Valenciana: districtualization, specialization and recent difficulties

Districtualization

The textile-clothing industry in the Comunidad Valenciana is defined by the high number of geographical agglomerations of firms, particularly in the provinces of Alicante and Valencia. Those agglomerations are either dependent on one large firm, completely dominate a village, or have been converted into local production systems. They have been studied to identify their main locations, to characterize the network organizational model or to determine the impact of concentration on firms (Ybarra, 1991; Soler, 2000). Those studies have agreed that the most important agglomerations of textile-clothing firms are located in the area of Alcoi-Ontinyent.[11] Together, they form the Valencia textile-clothing industrial district. Given the peculiar social and organizational characteristics of the area, that literature has agreed to consider it as an industrial district.

Specialization patterns

The main textile-clothing regions in Spain are specialized in different subsectors: 'clothing' in Galicia, 'textile fibres' and 'knitted and crocheted articles' in Cataluña, 'home textiles' and to some extent 'textile fibres' in the Comunidad Valenciana (INE, 2004). In 2000 the Comunidad Valenciana employed 16.6 per cent of the textile-clothing workers in Spain (Table 4.14), of which 67.2 per cent worked in textiles and 32.8 per cent in clothing (against 43.8 per cent and 56.2 per cent in Spain as a whole). Hence at the regional level there is a bias in favour of textiles compared to Spain as a whole. Within textiles, the main specialization occurs in textile fibres and home textiles (part of other textiles), with 20.9 per cent and 28.1 per cent of regional textile-clothing employment respectively. In particular, the region represents 33.7 per cent of the Spanish employment in home textiles, also above the textile-clothing regional share in employment (16.6 per cent).

The geographical distribution of the firm's subsectors of specialization confirms what has been said with respect to employment, with 64.5 per cent of the firms in the Comunidad Valenciana specialized in textiles compared to 38.9 per cent in Spain as a whole, and 35.5 per cent compared to 16.1 per cent in the case of home textiles (Table 4.15). Effectively, 33.4 per cent of Spanish home-textile firms are located in the Comunidad Valenciana. The specialization pattern of the Comunidad Valenciana is compared to that of Spain in Figure 4.10, which once again shows the relative concentration of home-textile firms in the region (33.4 per cent of the Spanish firms in this subsector).

Recent difficulties

The deterioration of the textile-clothing trade balance in the Comunidad Valenciana has followed a similar pattern to that of Spain and the EU-27 as a whole. The region had a trade surplus until 2004, incurring a deficit ever since (Figure 4.11). As for the Spanish or European cases, 2004 was a turning point for the textile-clothing industry in the region. However, unlike the previous cases, where the increase in imports was the main factor responsible for the aggravation of the already-existing trade deficit, in the case of Valencia the previously existing surplus was eliminated mainly by the large fall in exports that were not sufficiently competitive in third markets against new alternative suppliers in

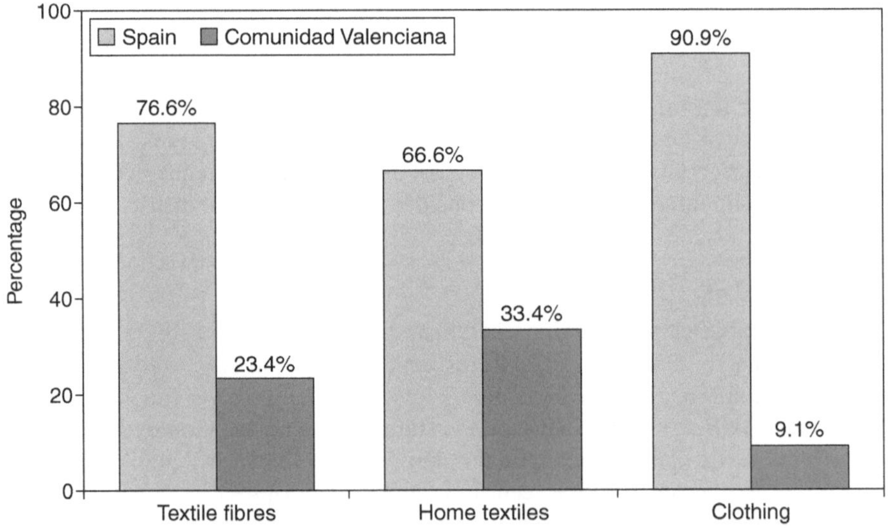

Figure 4.10 Distribution of firms by subsectors and location (source: constructed using data from IVE (2002)).

Table 4.15 Subsector structure of textile-clothing firms (2000)

Codes	*Description*	*Spain*		*Comunidad Valenciana*		
		Firms	*Share (%)*	*Firms*	*Share (%)*	*Regional share (%)*
17.1 17.2 17.3	Textile fibres	1,778	9.8	414	15.0	23.3
17.4	Made-up textile articles, except apparel	951	5.2	225	8.2	23.7
17.5	Other textiles	2,918	16.1	976	35.5	33.4
17.7	Knitted and crocheted articles	1,411	7.8	162	5.9	11.5
Textile total		7,058	38.9	1,777	64.5	25.2
18.2	Clothing	11,062	61.1	976	35.5	8.8
Total		18,120	100	2,753	100	15.2

Source: constructed using data from IVE (2002).

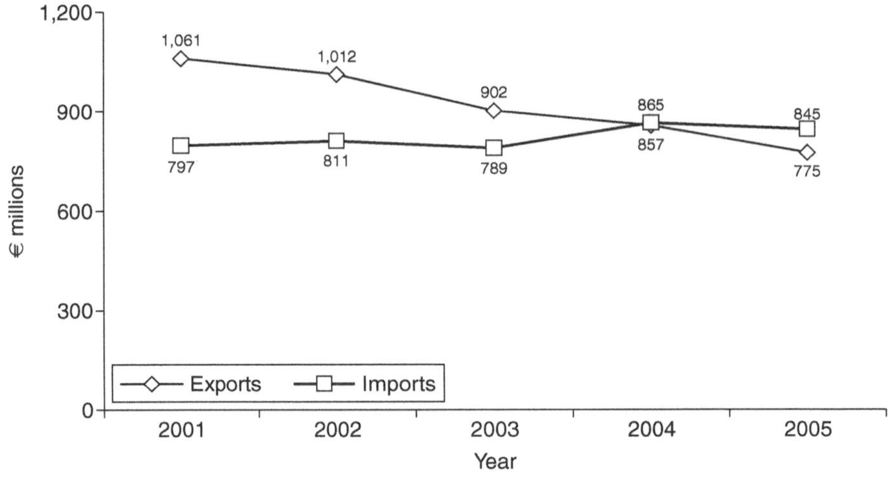

Figure 4.11 Textile-clothing external trade for the Comunidad Valenciana (€ millions, 2001–05) (source: constructed with data from CEUC (2006)).

Central and Eastern Europe and Asia (exports decreased by 27 per cent). On the other hand, imports increased by 6 per cent, with imports from Asia growing by 50 per cent (CEUC, 2006).

The adaptation to the new environment, with worldwide liberalization, European integration and falling domestic demand, has been reflected in increasing trade deficits, loss of production and employment and exit of firms. Figure 4.12 shows the evolution of those indicators for the textile-clothing sector in the Comunidad Valenciana and in Spain as a whole. In 2002–05, the export/import ratio fell by 26.5 per cent, production fell by 18.1 per cent, employment and the number of firms were reduced by 17.3 per cent and 12 per cent, respectively. These data, put together, illustrate the extent of the recent difficulties faced by the industry, but also show that the scenario in the Comunidad Valenciana was not as negative as in Spain as a whole.

4.4 Conclusions

The analysis presented in this chapter characterizes the textile-clothing sector in the EU-27, Spain and the Comunidad Valenciana and shows the challenges it faces in the context of globalization. A more detailed study of this impact will be the subject of a detailed survey study presented in the next two chapters. In the last decade, the processes of European integration and worldwide liberalization of the textile-clothing trade, together with changes in demand patterns in Europe, have forced the industry into decline and fostered a discussion on how best to cope with the new challenges. Some of the consequences have been plant closure and restructuring, market reorientation, and in some cases complete firm exit, with loss of employment and income (Table 4.16). Given the industry's tendency for strong regional agglomeration, the impact of employment and income loss in those regions that are very specialized in textiles-clothing is magnified and has the potential to create regional imbalances.

Moreover, the textile-clothing industry is representative of traditional manufacturing, with declining demand, growing competition and low technological intensity. The industry shows low levels of innovation, partly compensated by the diversification away from low value added subsectors, such as clothing, which are in decline, towards high value added subsectors, such as technical textiles, which show stronger growth. Hence, the textile-clothing sector is heterogeneous along the product and market dimensions.

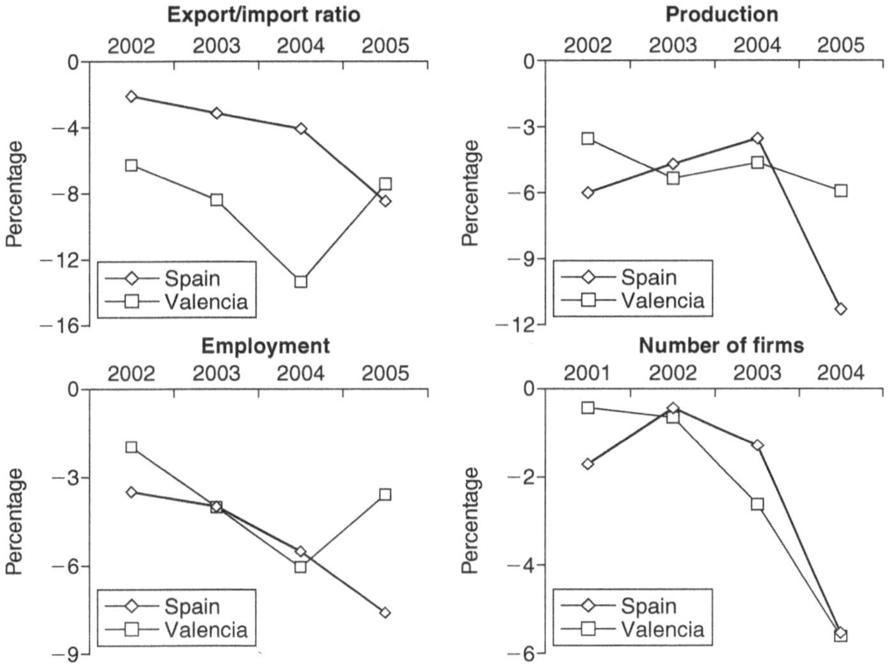

Figure 4.12 Evolution of textile-clothing indicators in the Comunidad Valenciana and Spain (%, 2002–05) (source: constructed using data from CEUC (2006) and INE (2005)).

Table 4.16 Summary of the evolution of the textile-clothing sector (%, 2001–05)

	EU	*Spain*	*Comunidad Valenciana*
Export/import ratio	–5.0	–16.7	–31.3
Production	–11.2	–23.4	–18.1
Employment	–15.2	–19.1	–17.3
Firms	–27.0	–13.0	–12.0

Finally, the textile-clothing industry is organized mainly around specialized SMEs that agglomerate geographically. In the case of a territory such as the Comunidad Valenciana, with a strong agglomeration of firms specialized in the home-textile subsector, there is the possibility of an uneven impact of globalization among those firms located in the Alcoi-Ontinyent industrial district and those outside, and among those firms specialized in the main subsector in the district and those with other specializations.

It has been shown that the territorial model of organization brings competitive advantages to the agglomerated firms and needs to be considered in an analysis of the restructuring strategies to be developed to overcome the current difficulties faced by the sector, not least because its evolution differs according to the territory and specialization patterns. In particular, the subsector that predominates in the Comunidad Valenciana, home textiles, shows a distinct evolution. This aspect is to be further investigated for several reasons. First, because firm performance will vary not only according to the district location or otherwise (district effect), but also according to the specialization pattern adopted (subsector effect). Second, differences in specialization in a district will also be responsible for differences in performance across districts. Finally, territorial characteristics will also have an impact on firm performance within the same district.

5 Methodology of analysis

An application to the Valencia textile-clothing industrial district

This chapter presents the territory that will be the basis of the case study presented in Chapter 6 (the Valencia textile-clothing industrial district) and establishes the methodology of analysis used to obtain homogeneous and reliable primary information from that territory. First, it describes the process of geographical delimitation of the industrial district as unit of analysis. Second, it explains the methodology of the study carried out in the Comunidad Valenciana through a questionnaire distributed to local district firms affiliated with the Association of Textile Entrepreneurs of the Comunidad Valenciana (ATEVAL), which is the Spanish textile-clothing association with the largest membership. Finally, it presents a preliminary analysis of the questionnaire results, which for now is essentially descriptive and will be developed further in Chapter 6.

5.1 Geographical definition of the industrial district

The search for instruments of analysis of industrial districts in advanced economies is made relevant today by the joint impact of the difficulties faced by traditional manufacturing in those economies, the potential for recovery of the organization model based on industrial districts and the emerging industrial policy with territorial roots. However, the use of different and complex methodologies hinders the appearance of a consensus regarding the methodology to be used in the geographical and productive delimitation of industrial districts and consequently calls into question the validity of the results obtained.

In fact, the lack of methodological definition has led some authors to consider the study of industrial districts as an eclectic field of research (Becchetti *et al.*, 2007). The inconsistencies found have their origins in data availability, researcher bias in favour of certain statistical criteria and the role of the procedure as an end in itself more than as a means for research (Brenner, 2003; O'Donoghue and Gleave, 2004). Independently of its reasons, to overcome the existing difficulties the chapter proposes an objective methodology applied to the Spanish textile-clothing industry. To that end, the chapter starts by laying out the main theoretical perspectives on the geographical delimitation of the industrial district in order to obtain an objective methodology. The second part of the chapter applies that methodology to the identification of the agglomerations that can be found in a particular industry (textile-clothing) in a particular European country (Spain).

5.1.1 Perspectives on the geographical delimitation of the industrial district

The territorial analysis of industrial districts has followed three major lines of investigation: (1) the identification of geographical areas with industrial district characteristics; (2) the study of the characteristics of the industrial districts previously identified; and (3) the integration of both strands (identification and validation).

Geographical delimitation

The most important efforts to identify geographical areas that could be classified as industrial districts have been carried out by Italian researchers, such as Sforzi (1990), or the Italian National Institute of Statistics (ISTAT). Sforzi has identified Italian industrial districts using the notion of local labour markets (LLM), defined as regional commuting areas. For Italy, he identified 955 LLM,[1] which were further divided into 15 groups of local systems. Only one of those groups has the characteristics of the light industrialization model[2] and that group was used to analyse the economic structure of this type of local production system. In the group, the areas that are specialized in one industry are identified as industrial districts. Sforzi found 61 districts, located mainly in the northwest and centre of Italy, and specialized mainly in traditional manufacturing (textile-clothing, footwear, leather, wooden furniture, metal, pottery, musical instruments and toys).

Following the work of Sforzi, there have been other studies on Italy and other countries such as France, Spain and Portugal, which try to define with greater precision the location of industrial districts from a quantitative perspective.[3] In particular, ISTAT's (1996) study of Italy is based on the idea that, for an agglomeration of firms to be regarded as an industrial district, it must meet certain sectoral specialization criteria with respect to employment shares:

- share of manufacturing employment above the national average;
- share of SME employment above the national average;
- share of employment in the main industry of the territory above the national average;
- share of SME employment in that industry above the national average.

Similarly to Sforzi (1990), ISTAT identified the specialization of most industrial districts as being in mature labour-intensive industries such as textile-clothing, pottery and furniture. The ISTAT methodology was also used by Boix and Galletto (2005) for Spain, Hernández *et al.* (2005) for Cataluña and Lainé

(2000) for France. This author defined territory as the area of employment – that is, the geographical space where most workers live.[4] Different methodologies have been applied by Cerejeira (2002) for Portugal and Courlet and Pecqueur (1994) for France. Cerejeira (2002) identifies the Portuguese industrial districts using 16 spatial units at the municipality level (*concelhos*), which have identical characteristics to those defined by Sforzi (1990). The industrial districts are the spatial units that meet four characteristics implicit in the research on industrial districts:

- The industrial profile of the *concelho* and its specialization pattern, measured as the share of manufacturing employment in the *concelho.*
- The concentration of employment in SMEs, measured as the number of firms with less than 50 employees per square kilometre in the *concelho.*
- The spatial agglomeration of firms with respect to the national average, measured as the share of industrial firms in the *concelho* divided by the national average.
- The geographical agglomeration of firms and the specialization level of the *concelho*, measured as the number of industrial firms in the main sector of the *concelho* per square kilometre.

Cerejeira (2002) concluded that 13 per cent of Portuguese manufacturing employment is located in industrial districts, specialized in traditional manufacturing (textile-clothing, wooden furniture, glass, metallic products, pottery and leather).

The work of Courlet and Pecqueur (1994) was based, similarly to Lainé (2000), on INSEE data from the *Annual Company Survey for France* in 1985. They found 44 employment areas with industrial district characteristics, using a methodology based on LLMs in particular industries, which defines a number of criteria:

- existence of an industry with more than 10 per cent of employment;
- that industry should have at least 5 per cent of its national employment in the district;
- the industry should have at least ten firms with more than ten employees.

Courlet and Pecqueur (1994) found that most activities in these circumstances are within the textile-clothing industry.

The studies presented thus far reach similar conclusions with respect to the importance of the presence of the textile-clothing industry in industrial districts in Italy, France, Portugal and Spain.

Looking more closely at the case of Spain, authors such as Costa (1988) for the Catalunian Vallés Oriental region and Ybarra (1991) for the Comunidad Valenciana have established a methodology that has been widely used in subsequent studies. Ybarra (1991) assumes that industrial districts specialize in specific sectors and are composed of SMEs. Using the *National Company Survey*, three steps were followed:

1 identification of economic activities that are more characteristic and/or more dynamic at the regional level than at the national level;
2 location of those activities at the municipality level;
3 identification of SMEs in those municipalities and activities.

In the first step, Ybarra identifies the economic activities with a higher weight in the region than in the country with respect to employment, investment and the number of new firms. The regional weight with respect to the national average is calculated using the coefficient of specialization (CS)[5] and 18 out of 31 sectors are found to be more present in the Comunidad Valenciana than in Spain as a whole. In the second step, a location procedure is followed for those 18 sectors: (1) the Herfindahl Index (HI)[6] is calculated for employment, investment and new firms in each sector at the local level; (2) the coefficients of localization (CL)[7] are calculated for all municipalities; (3) the municipality share in the sector is calculated with respect to total sector activity. Putting these measures together, it is possible to identify the municipalities at the centre of the district and those in its area of expansion. From this a total of 11 larger districts and six smaller districts is obtained. Finally, the third step acknowledges that SMEs are the fundamental production agents responsible for most economic activities representative of industrial districts.

Ybarra suggested that other variables of qualitative nature would also be helpful in characterizing the industrial district:

- the technical characteristics specific to an activity, sector or product;
- the social, work and cultural environment of an activity;
- the administrative and political conditioning of an activity and territory.

In conclusion, from the perspective of identification of industrial districts, the objective of the researcher is to pinpoint the existence of territories that can be classified as industrial districts in a particular country. Although the methodology followed by the various studies referred to here differs according to the data and the statistical measures used, they all find that traditional manufacturing in Europe tends to concentrate geographically, with some of those concentrations possessing the characteristics of an industrial district.

Characteristics of the districts

Another strand of research verifies whether an industrial agglomeration can be classified as an industrial district according to the following criteria:

- the existence of many SMEs and microfirms in the territory;
- the territory is a small area that is specialized in a particular sector;
- there is a high interrelation between the district as a production unit and the territory as a social area.

The methodology followed by this strand consists of selecting the local production system to analyse, defining the municipalities with a higher share of the production activities to be studied, and finally using that information to find the social and organizational characteristics of the district in the territory. Taking the example of the ceramic industrial district in the Spanish province of Castellón, Martínez (2001) and Molina (2002) studied the location of firms in the municipalities of l'Alcatén, Plana Baixa and Plana Alta. The sample was then restricted to the two most important subsectors (pavement and tiles), concluding that those municipalities are very specialized in ceramic production and that production is strongly interrelated with the social territory. The area was thus identified as the ceramic industrial district of Castellón.

An integrative proposal

It is possible to combine the identification of the geographical area of the district and the study of its characteristics, be it organizational, strategic or competitive. Taking the example of the textile-clothing sector in the Comunidad Valenciana, Santa María (1999) extended the methodology of Ybarra (1991) to consider two steps in the identification of the industrial district, the first one being more quantitative and the second being more qualitative. The first stage defines which activities are central or complementary to a sector and what their location patterns in the territory are. In the second stage, the study focuses on the agglomerations, which are studied with respect to their structure and functioning: number of employees, turnover, market positioning, specialization, distribution channels, propensity to export and decentralization of production. The industrial district characteristics that have been identified are: the presence of many SMEs in the territory; the specialization of those firms in one single sector; and the strong interrelation between business and social activity in the territory.

Also for the case of the Comunidad Valenciana, Soler (2000) identified the main industrial agglomerations in each sector and the respective degree of specialization as necessary conditions for the existence of a district. The next step was the investigation of the presence of structural characteristics of an industrial district in those agglomerations and of the origin of external economies in the territory (locational economies or urbanization economies).[8] The agglomerations were defined on the basis of the HI and the CS for firm-level employment data provided by the SME Institute of the Ministry of Industry for 1992.[9] It was found that 90 per cent of the employment in the Comunidad Valenciana originates in 12 sectors, out of which four (textile-clothing, footwear, ceramics and furniture) meet the criteria to be considered as organized in industrial districts, with textile-clothing and footwear presenting the highest levels of specialization.

The perspectives summarized in Table 5.1 are a departure point for the establishment of a methodology. The contributions of Santa María (1999) and Soler (2000) will be taken as the basis of the next section.

Table 5.1 Perspectives on the geographical delimitation of the industrial district

Author	*Objective*	*Methodology*
• Boix and Galletto (2005) • Cerejeira (2002) • Courlet and Pecqueur (1994) • ISTAT (1996) • Lainé (2000) • Sforzi (1992) • Ybarra (1991)	Identification of areas that can be classified as industrial districts	1 Identification of production systems 2 Statistical analysis of their characteristics 3 Classification as an industrial district if the criteria are met
• Martínez (2001) • Molina (2002)	Study of the characteristics of areas classified as districts	1 Geographical identification of the production system 2 Delimitation of production activities 3 Identification of the main locations 4 Verification of the industrial district characteristics
• Santa María (1999) • Soler (2000)	Identification and study of the industrial district	1 Identification of production activities 2 Statistical identification of locations 3 Geographical location of district areas 4 Empirical validation of the existence of a district

5.1.2 The identification of textile-clothing industrial districts in Spain

Following on the methodology previously described, the first step towards the identification of textile-clothing industrial districts in Spain is the choice of database to use in the statistical analysis and the choice of the textile-clothing activities to include. The second step is the construction of a statistical measure, such as the CS, in order to identify the main textile-clothing industry agglomerations. Finally, it is verified whether those agglomerations meet the criteria to be classified as industrial districts.

The methodology used here improves the work of the previous authors in two important aspects. First, it explicitly recognizes which production activities can be regarded as district activities and, second, it uses a more detailed database than the ones used previously. The use of a detailed database made available in recent years allows the detection of new textile-clothing geographical concentrations that can be classified as industrial districts. The steps followed are:

1 choice of database;
2 delimitation of the textile-clothing activities to be included;
3 statistical measurement of geographical concentration using CS;
4 territorial identification of industrial districts.

Choice of database

The Spanish National Institute of Statistics (INE) made available its latest National Census data in 2004 (INE, 2004).[10] The data are based on individual information gathered in individual interviews. From this large database, information was extracted on the number of employees over 16 years old, classified by economic activity (three-digit CNAE-93) and municipality.

Delimitation of the textile-clothing activities to be included

As before, only the textile industry (code 17) and the clothing industry in a strict sense (code 18.2) will be included in the analysis that follows. The inclusion of these two codes facilitates the statistical handling of the data and allows the inclusion of very small villages for which the level of statistical disaggregation is low due to reasons of confidentiality.

Statistical measurement of the main textile-clothing concentrations

The importance of the textile-clothing industry in each territory is quantified using the CS, which defines the level of industrial concentration in each region, province or municipality with respect to the national average. The most important textile-clothing concentrations in Spain as defined in Table 4.12 are, at the regional (and provincial) levels, Castilla-La Mancha (Toledo, Ciudad Real, Albacete, Cuenca, Guadalajara), Cataluña (Barcelona, Girona, Lleida,

Tarragona), Comunidad Valenciana (Alicante, Valencia, Castellón) and Galicia (Coruña, Ourense, Pontevedra, Lugo). In this chapter, the municipalities with higher CS within those selected provinces will be identified in order to find those geographical concentrations that can be classified as industrial districts.

Territorial identification of industrial districts

For a municipality to be included in the analysis it must meet four criteria. It must:

1 have more than 500 employees (smaller municipalities are subject to confidentiality issues);
2 have at least ten textile-clothing firms to ensure a minimum level of diversification;[11]
3 have a CS higher than 10;
4 be contiguous to at least another of the municipalities included (otherwise it is excluded even if meeting the other three criteria).

In the whole of Spain, there are only 38 municipalities that meet the first three criteria (Table 5.2).[12] The last criterion of contiguity was investigated using the mapping of all 38 municipalities above, as shown in Figure 5.1. It was found that Mataró and Sonseca are disconnected from the other 36 and for that reason were eliminated. This leaves two geographical concentrations that meet the four criteria: the municipalities located around Igualada and the municipalities located around Alcoi and Ontinyent. For comparison, Table 5.3 summarizes the

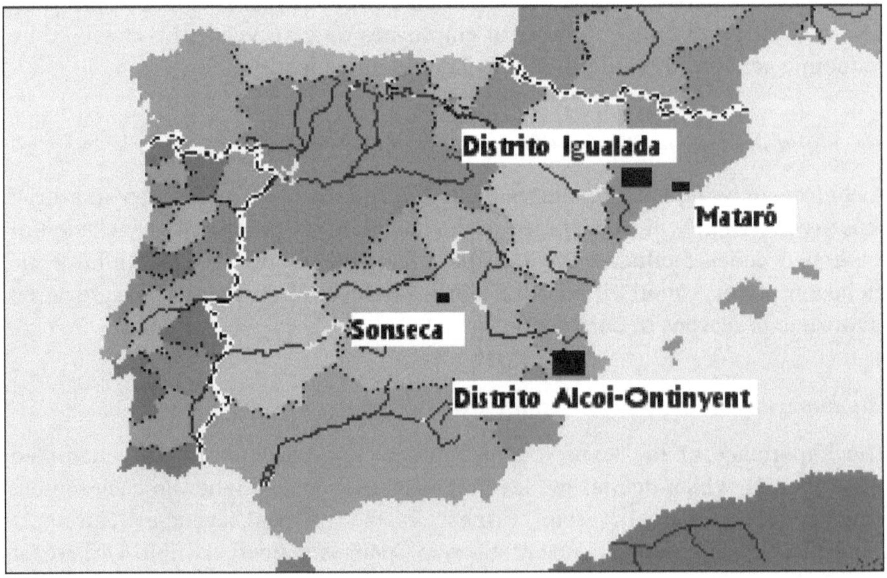

Figure 5.1 Map of the main textile-clothing concentrations in Spain.

information about the four areas represented in Figure 5.1. Among these, the area of Alcoi-Ontinyent has both the largest number of employees and the largest CS. For this reason, it was chosen as the main case of analysis and it is believed to be representative.

The four areas identified all have a high number of SMEs, are dedicated to a dominant industry and, due to the territorial networks established over time,

Table 5.2 Spanish municipalities with highest CS in the textile-clothing industry

Municipality	*Employees*	*CS*
Villafranca del Cid (CS)	574	29.98
Banyeres de Mariola (A)	1,304	28.25
Sant Bartomeu del Grau (B)	249	28.15
Agullent (V)	444	27.52
Albaida (V)	1,012	24.45
Bocairent (V)	627	19.96
Ontinyent (V)	4,407	19.30
Prats de Lluçanès (B)	422	19.10
Atzeneta d´Albaida (V)	163	18.90
Vilanova del Camí (B)	1,499	18.56
Muro de Alcoy (A)	960	18.08
Cocentaina (A)	1,313	17.99
Olost (B)	154	17.15
Hostalric (GI)	378	15.73
S. Margarida de Montbui (B)	982	14.01
Òdena (B)	291	13.77
Sonseca (TO)	1,000	13.60
Alcoy/Alcoi (A)	5,071	12.75
Montaverner (V)	162	12.68
Canals (V)	1,095	12.56
Alcúdia Crespins (V)	345	12.51
Aielo de Malferit (V)	344	12.44
Navarcles (B)	504	11.94
Pobla de Lillet (B)	105	11.78
Igualada (B)	3,036	11.67
Avinyó (B)	183	11.51
Beneixama (A)	145	11.49
Velada (TO)	177	11.48
Bihar (A)	268	11.31
Mataró (B)	8,668	10.86
Mediona (B)	122	10.83
Alfarrasí (V)	93	10.78
Tordera (B)	795	10.76
Anglès (B)	396	10.74
Crevillent (A)	1,669	10.61
Cercs (B)	94	10.61
Belvís de la Jara (TO)	84	10.48
S. Pablo Montes (TO)	142	10.31

Note
(A) Alicante; (B) Barcelona; (CS) Castellón; (GI) Girona; (TO) Toledo; (V) Valencia.

there is a strong interrelation between production activities and the social life of the territory. In addition to these four areas, Boix and Galletto (2005), due to their more general approach, have identified 50 others. This more general approach included: defining the textile-clothing industry by including subsectors that have been excluded here; defining a lower CS threshold; and taking one district per municipality. The more stringent criteria used here allow the emergence of only four areas, two of which can strongly be regarded as industrial districts, and allows the straightforward identification of the most representative textile-clothing industrial district (Alcoi-Ontinyent).

In conclusion, the methodology of geographical and productive delimitation of industrial districts proposed here for the Spanish textile-clothing industry shows that, similarly to other traditional manufacturing in other European countries, there is in the particular case studied a strong tendency towards agglomeration. The more structured approach followed presents some advantages. On one hand, it is easy to transpose to other traditional manufacturing industries by simply selecting the subsectors of activity to be considered and defining the thresholds regarding the characteristics of the industry. On the other hand, by choosing a level of analysis at a higher spatial scale (municipal instead of provincial or regional) it is sufficient that a firm is located in the relevant municipalities to be considered as a district firm.

Table 5.3 Employment and specialization in the main textile-clothing concentrations in Spain

	Employees	*CS*
Municipalities in the Toledo province		
Sonseca	1,000	13.60
Municipalities in the Barcelona province		
Mataró	8,668	10.86
Distrito Igualada (Barcelona province)		
Vilanova del Camí	1,499	18.56
Santa Margarida de Montbui	982	14.01
Òdena	291	13.77
Igualada	3,036	11.67
Total	5,808	13.47
Distrito Alcoi-Ontinyent (Valencia province)		
Banyeres de Mariola	1,304	28.25
Agullent	444	27.52
Albaida	1,012	24.45
Bocairent	627	19.96
Ontinyent	4,407	19.30
Muro de Alcoi	960	18.08
Cocentaina	1,313	17.99
Alcoi	5,071	12.75
Total	15,138	17.10

5.2 Methodology of analysis

5.2.1 Population and sample

The population of the study is formed by all textile-clothing firms in Spain. When the questionnaire was distributed in 2005, there were 15,747 firms registered (INE, 2005). Since it was found that the most important textile-clothing district was located in the Comunidad Valenciana area of Alcoi-Ontinyent, the database used to build the sample was obtained from ATEVAL.[13] This association is one of the most important textile-clothing associations in Spain, representing around 10 per cent of the industry – that is, 570 textile-clothing firms with around 19,000 employees, located mainly in the Comunidad Valenciana, but also including firms from Cataluña and the Castillas, and specialized in all the various textile-clothing subsectors previously identified.

Accordingly, the analysis that follows will focus on the Alcoi-Ontinyent industrial district. It will take into account the heavy specialization of the Comunidad Valenciana in home textiles, defining the subsectors as follows:

- *Textile fibres*, which groups firms with their main activity in subsectors 17.1, 17.2 and 17.3 (spinning, weaving and finishing).
- *Home textiles*, which groups firms with their main activity in subsectors 17.4 and 17.5 (home and technical).
- *Clothing*, which groups firms with their main activity in subsectors 17.6, 17.7 and 18.2 (knitted, crocheted and wearing apparel).
- *Other activities*, a residual category that includes all firms listed by ATEVAL whose main activity cannot be included in any of the previous categories (for example, plastic containers, dyes, etc).

5.2.2 Questionnaire

The required information was collected by means of a questionnaire reproduced in Appendix 1, which was distributed to the firms listed by ATEVAL. The questionnaire was structured taking into account both the characteristics of the sampled firms and the objectives of the study, meeting the criteria of simplicity, brevity, precision and relevance. To facilitate the responses, the following criteria were followed:

1 The questionnaire was easy to complete, with short, unambiguous questions and no requests for confidential information. Complementary information such as the number of employees, year of establishment and main activity, was taken from the AMADEUS database and the firms' records in the archives of ATEVAL.
2 The questions were, in most cases, closed or semi-open to facilitate comparisons and clear conclusions.
3 Immediate return was made possible by including a self-addressed stamped envelope and providing a fax number.

The questionnaire was structured around five sections:

1 *Identification of the firm, respondent, competitive situation and international activity*: this section provides information about the sociocultural characteristics of the respondent and respective activity in the firm, the respondent's perceptions of the current and future competitive situation of the firm and the current internationalization strategy of the firm.
2 *Strategic factors*: this section asks the respondent to make subjective evaluations on the impact of strategic factors present in the general and specific environment on the firm's performance.
3 *Determinants of competitiveness*: this section gathers comparative information and evaluates the competitive position of the firm vis-à-vis its main competitors with respect to the proposed criteria.
4 *Business strategy*: this section analyses the need felt for each of the proposed strategies.
5 *Public policies*: this section evaluates the urgency of the main demands made by textile-clothing firms to policy-makers.

The questions that compose the questionnaire use the following procedure:

1 Compilation of general information about the current state of the textile-clothing industry using the press, industry publications, statistical sources such as AMADEUS, etc.
2 Face-to-face semi-structured interviews with 50 managers and local government officials.
3 Use of two panels of experts to synthesize, prioritize and operationalize key aspects of the information obtained in the first two stages. One panel was formed by business agents (owners, managers and representatives of industry associations). Another was made up of territorial agents (local development agents, policy-makers and job-training centres).

The next step was the definition of the scales and items, which varied according to five different situations:

1 For the profile of the respondent, perceptions and the firm's internationalization strategy, both closed and open questions were used.
2 For the *strategic factors* the scale was formed by 16 items that captured the diversity of environmental threats and opportunities. A five-point Likert scale was used to gather the general opinions of the experts. According to that scale, the respondent evaluated the influence of those factors using values from 1 (affects very negatively) to 5 (affects very positively).
3 For the *competitive position*, the five-point Likert scale was formed by 15 items. Each respondent had to evaluate the position of the firm vis-à-vis its main competitors at home and abroad.
4 For the *business strategies*, the five-point Likert scale was formed by 13 items by which to judge the relevance to the firm's survival using values

from 1 (not necessary) to 5 (very necessary). In the end, the various strategies were ranked with respect to their perceived importance.

5 For *public policies*, 11 items suggested by the literature and experts were valued in terms of their priority using values from 1 (not needed) to 5 (urgently needed).

5.2.3 Field work

Before sending the questionnaire to the full sample, it was tested among five people close to the industry (entrepreneurs, managers and industry association personnel). Their suggestions allowed the improvement of the questionnaire. Its final version was posted in May 2005, together with a presentation letter signed by ATEVAL, to all sampled firms according to ATEVAL records. Following a second round for lagging respondents in September 2005, a total of 207 responses were obtained. Among these, six were rejected (three due to important errors in their completion, two were void and one was duplicate). To ensure that the early respondents and late respondents did not systematically differ, these two groups were compared through independent sample t-tests to check for equality of means. The analysis indicated no significant difference in the variables of interest between late and early respondents. Additionally, the responding firms were compared to a random sample of 20 non-respondents regarding size (sales volume) and experience (years since foundation). No significant differences were found ($p < 0.05$), providing no evidence for non-response bias. Finally, those variables from the survey responses were cross-checked against company reports and published data, where possible. A high degree of correspondence between published data and survey responses was found, supporting the veracity of the survey responses.

This procedure left 201 valid questionnaires out of 570 firms listed by ATEVAL, representing a 35 per cent response rate, a very high value that corresponds to over 7,000 textile-clothing employees (4 per cent of the Spanish total). The sample error is ±5.7 per cent for a 95 per cent confidence level.[14] Box 5.1 summarizes the methodology.

Box 5.1 Summary of the methodology

Domain: Spain.
Universe: 570 textile-clothing firms listed by ATEVAL.
Sample size: 201 firms completed the questionnaire correctly (response rate above 35%).
Sample error: ±5.7% (5% significance level).
Dates: Between May and September 2005.
Methodology: Questionnaire with 70 semi-closed questions posted to all firms listed by ATEVAL in May 2005. Enclosed was a letter from ATEVAL urging firms to respond. The questionnaire was re-sent to lagging respondents in September 2005. In the few cases where posting was not viable, fax was used instead. Similarly, for small errors or blanks the respondent was contacted again for clarification.

5.3 Organizational characteristics of the sample

The validity of the sample will be checked by means of a preliminary descriptive analysis. In particular, the 201 firms that returned valid questionnaire responses will first be characterized with respect to their size, age, location, subsector of specialization and internationalization activities. In a second stage, the relationships between those variables will be studied with respect to their degree, sign and significance.

5.3.1 Descriptive analysis

The descriptive analysis of the sample exposes the distribution of the values taken by the variables that characterize the sample firms, summarized by a set of descriptive statistics presented in tables and graphs.

Starting with the geographical location of the sample firms (Figure 5.2), 91 per cent (183 firms) are located in the Comunidad Valenciana, out of which 48 per cent are located in the province of Valencia, 38 per cent in the province of Alicante and 5 per cent in the province of Castellón. The remaining 9 per cent (18 firms) are located in other Spanish regions (by province, 12 in Barcelona, two in Girona, two in Madrid, one in Murcia and one in Valladolid). In the sample, 56.2 per cent of the firms (113 firms) are located in the Alcoi-Ontinyent industrial district and 43.8 per cent (88 firms) are located outside (Figure 5.3).

At the municipality level, Table 5.4 shows that only one of the municipalities with the largest number of sample firms (Crevillent) is outside the Alcoi-Ontinyent industrial district. All eight municipalities forming the district are those with the largest representation in the sample. Another 18.4 per cent of the firms locate in municipalities with more than one sample firm, with the remaining 19.4 per cent locating in 39 municipalities with one sample firm each.

In the sample, 23.9 per cent are microfirms, 55.7 per cent are small firms, 16.9 per cent are medium firms and only 3.5 per cent are large firms (Figure 5.4).[15] The sample structure should be compared to the average structure for all Spanish textile-clothing firms in 2000–04 (INE, 2005), of which 73.2 per cent

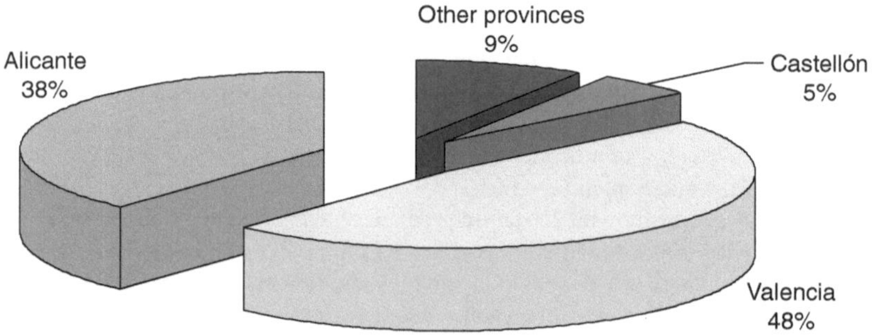

Figure 5.2 Provincial distribution of the sample firms.

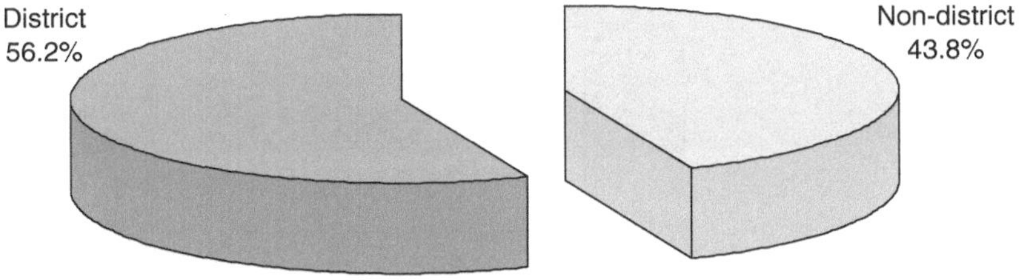

Figure 5.3 Locational distribution.

were microfirms, 23.4 per cent were small firms and only 3.4 per cent were medium and large firms. Hence, the interpretation of the sample results should take into account the over-representation of small and medium firms, as well as the under-representation of microfirms.

With respect to the age of the firms, 52.5 per cent of the sample is formed by firms that have been in business for 11–30 years (Figure 5.5). In the absence of data for the age of textile-clothing firms in the whole of Spain, the adjustment of the sample to the population can only be measured indirectly. Given that the literature has shown that age and size are highly correlated, one can make use of this relationship to infer the sample adjustment for the age variable. The correlation coefficient (Spearman's Rho) between age and size for the sample is 0.407 and significant at the 1 per cent level. Taking into account that firm size was biased upwards in the sample, there are reasons to believe that the same is true for firm age – that is, the sample firms may be on average older than the population of Spanish textile-clothing firms. This information should be taken into account when interpreting the results.

The subsector structure of the sample (Figure 5.6) shows that 45.2 per cent of the firms produce home textiles, 28.1 per cent produce textile fibres, 18.1 per cent produce clothing and only 8.5 per cent of the sampled firms have ancillary activities. The sample distribution should be compared to the population distribution. In

Table 5.4 Most represented municipalities in the sample

Municipality	*Number of firms*	*%*
Ontinyent	31	15.4
Banyeres de Mariola	17	8.5
Alcoi	14	7.0
Cocentaina	13	6.5
Muro de Alcoi	12	6.0
Crevillent	11	5.5
Albaida	10	5.0
Bocairent	9	4.5
Agullent	8	4.0
Total	125	62.2

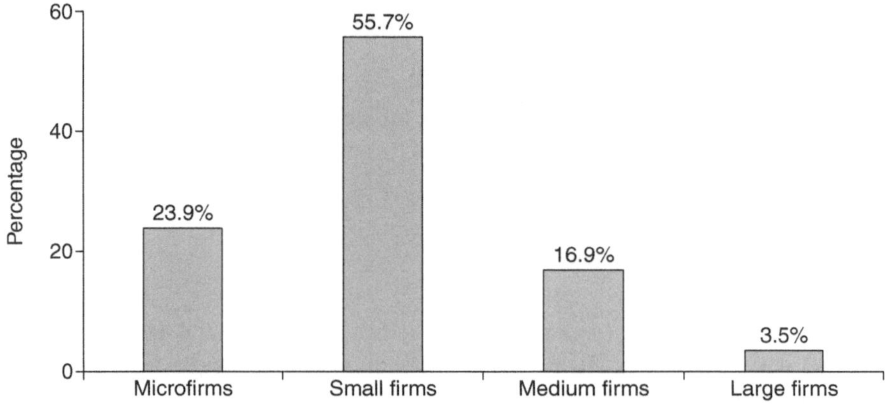

Figure 5.4 Distribution of firm size according to the number of employees.

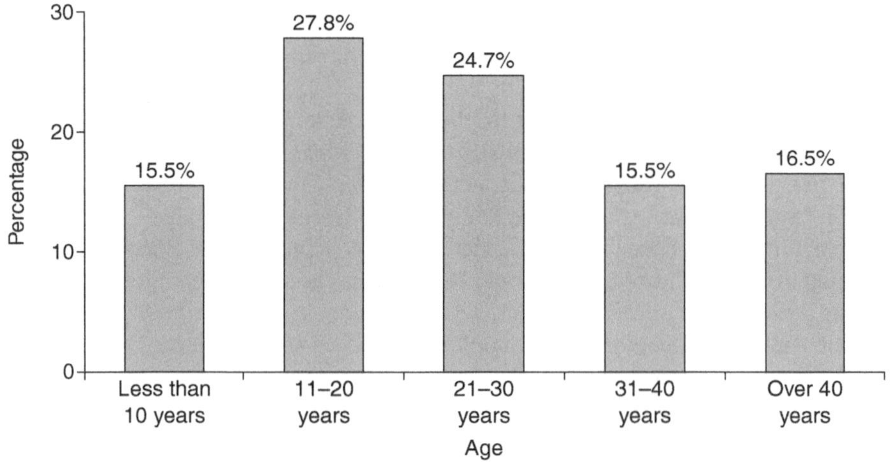

Figure 5.5 Age distribution of the sample.

the Comunidad Valenciana in 2000, 35.5 per cent of the textile-clothing firms produced home textiles, 23.2 per cent produced textile fibres and 41.3 per cent produced clothing (IVE, 2002). Hence, the sample over-represents home textiles and under-represents clothing. One explanation for this difference is that most clothing firms in the Comunidad Valenciana are microfirms with low participation in industry associations and are very dispersed geographically, whereas home-textile firms show stronger participation in industry associations, are larger and concentrate to a large extent in the Alcoi-Ontinyent industrial district (Golf *et al.*, 2005). The specialization structure can also explain the sample bias with respect to size and age, given that the home-textile firms which are over-represented also tend to be

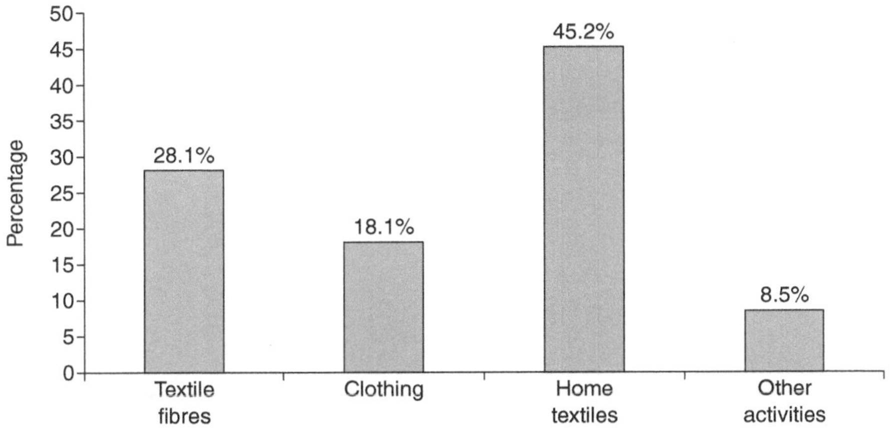

Figure 5.6 Subsector structure of the sample.

larger and older firms. This is an important characteristic of the sample that has to be taken into account in the analysis.

Finally, the bulk of the sample firms (94.5 per cent) have followed some form of internationalization, with 63.3 per cent both exporting and importing, 16.4 per cent only exporting, and 14.8 per cent only importing (Figure 5.7). However, two caveats must be taken into account when interpreting the results: (1) only 63.7 per cent of the respondents answered the question on internationalization; and (2) the exporting and/or importing intensity is not known, and some firms could export and/or import only sporadically.

In summary, the descriptive analysis of the sample has revealed little difference between the population of textile-clothing firms in the Comunidad Valenciana and in Spain as a whole. Not only are the sample firms larger and older than average, but home textiles are over-represented and clothing is under-represented in the sample. Moreover, the internationalization variable refers only to the subset of firms that answered the question. These results confirm the validity of the sample, at least with respect to its structural characteristics, taking into account the limitations previously described.

5.3.2 Relationships between the sample characteristics

The sample is also characterized by qualitative variables, which demands the use of contingency tables to analyse the relationships between those variables.[16] The statistical significance of the relationship is given by a Chi-square test with the null hypothesis of independence (no relationship). Fixing a confidence level of 95 per cent, p-values lower than 0.05 imply the rejection of the null hypothesis, hence the acceptance of the existence of a statistically significant relationship between the pair of variables under consideration.

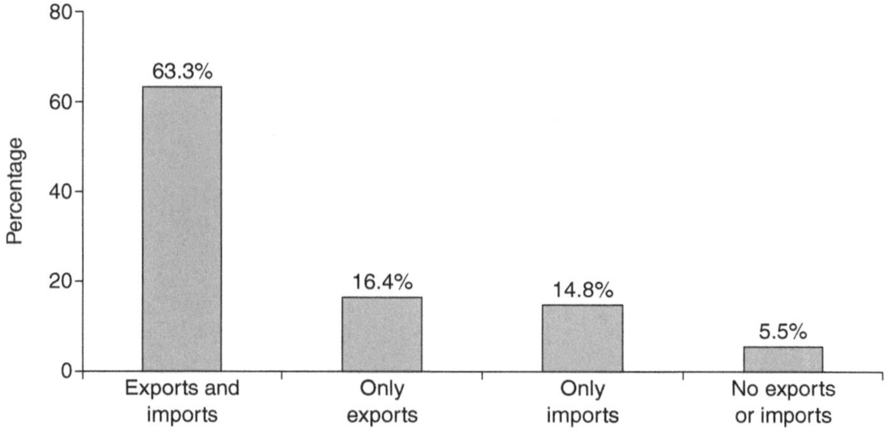

Figure 5.7 International activity of the sample firms.

In the sample there is no statistical relationship between district location and firm size or age (Tables 5.5 and 5.6), given the symmetric distribution of the corrected residuals[17] and the lack of significance of the Chi-square test. Hence, in the sample firm size and firm age are independent of locating in a district or not.

District location is, however, significantly associated with the subsector of specialization (Table 5.7). The result is due to the higher than expected location of clothing firms outside the district and of home-textile firms inside the dis-

Table 5.5 Association between firm size and location

	Non-district	*District*
Microfirm	0.7	–0.7
Small firm	–0.6	0.6
Medium firm	0.0	0.0
Large firm	–0.1	0.1

Note
Chi-square's p-value=0.922

Table 5.6 Association between firm age and location

	Non-district	*District*
0–10 years	–1.5	1.5
11–20 years	0.6	–0.6
21–30 years	–0.9	0.9
31–40 years	–1.1	1.1

Note
Chi-square's p-value=0.300.

trict, as given by the two significant cells in the table.[18] This observation is in line with the descriptive observation that the sample over-represents home-textile firms, which make up 45.2 per cent of the sample, and this happens because the presence of home-textile firms in the district is above what would be expected.

The association between firm size and subsector is not statistically significant, although large firms are more present among textile fibres producers (Table 5.8). On the contrary, the age distribution of firms is not related to their subsector in any of the cases (Table 5.9).

Internationalization is strongly associated with firm size (Table 5.10). More small firms are absent from external markets than expected and more medium and large firms simultaneously export and import than would be expected. The importance of firm size for internationalization is well documented in the economics and business literature and has already been discussed in the book. The relationship can be explained, among other reasons, by the fixed costs that firms must incur to enter a new market, which become bearable only after a certain capacity threshold, implying a minimum dimension required for international-

Table 5.7 Association between location and subsector

	Textile fibres	*Clothing*	*Home textiles*	*Other activities*
Non-district	–0.4	*3.9	–2.3	–0.7
District	0.4	–3.9	*2.3	0.7

Notes
* significant at 5%; Chi-square's p-value=0.01.

Table 5.8 Association between firm size and subsector

	Textile fibres	*Clothing*	*Home textiles*	*Other activities*
Microfirm	0.0	0.3	–0.9	1.2
Small firm	–1.1	0.6	0.7	–0.3
Medium firm	0.6	–1.1	0.6	–0.6
Large firm	*1.7	–0.3	–0.9	–0.8

Notes
* significant at 10%; Chi-square's p-value=0.658.

Table 5.9 Association between firm age and subsector

	Textile fibres	*Clothing*	*Home textiles*	*Other activities*
0–10 years	0.8	0.4	–1.6	1.0
11–20 years	–0.6	0.6	0.3	–0.4
21–30 years	–0.4	0.7	–0.1	–0.1
31–40 years	0.8	–1.2	0.8	–1.1

Note
Chi-square's p-value=0.848.

ization. Although this constraint has been removed to a large extent for information-based activities, it still remains for manufacturing firms.[19]

Internationalization is not, on the whole, associated with firm age (Table 5.11). Although the business literature has highlighted that firms take ever less time to start internationalizing, there are more young firms (less than ten years of activity) not engaging in international activity and more old firms (more than 40 years of activity) engaging in international activity than would be expected. The reduction in the start of internationalization lag is seen particularly in the case of high-tech and information-based firms. In traditional manufacturing, firms still tend to wait for some time before they start internationalizing, probably because they incur higher fixed costs of entry compared to knowledge-based activities.

There is no association at any level between district location and international activity (Table 5.12), but there is, however, a significant association between subsector and international activity (Table 5.13). In particular, less textile fibres

Table 5.10 Association between firm size and international activity

	Exports and imports	*Only exports*	*Only imports*	*No exports or imports*
Microfirm	–1.6	1.4	0.4	0.5
Small firm	–2.0	0.2	0.9	*1.8
Medium firm	**3.2	–1.2	–1.3	–2.3
Large firm	*1.9	–1.1	–0.6	–1.0

Notes
** significant at 5%; * significant at 10%; Chi-square's p-value=0.049.

Table 5.11 Association between firm age and international activity

	Exports and imports	*Only exports*	*Only imports*	*No exports or imports*
0–10 years	–1.7	–0.3	0.9	**2.1
11–20 years	–0.9	–0.1	–0.1	1.5
21–30 years	–0.8	1.4	1.4	–1.2
31–40 years	1.5	–0.1	–1.1	–1.2
Over 40 years	*2.3	–0.9	–1.2	–1.4

Notes
* significant at 5%; Chi-square's p-value=0.137.

Table 5.12 Association between location and international activity

	Exports and imports	*Only exports*	*Only imports*	*No exports or imports*
Non-district	0.2	–1.5	0.7	0.8
District	–0.2	1.5	–0.7	–0.8

Note
Chi-square's p-value=0.376

producers have internationalized, more clothing firms are importers and more home-textile firms have fully internationalized than would be expected. Hence, the home-textiles subsector is the most internationalized and the textile fibres subsector is the most domestically oriented. This is not surprising, as textile fibres producers are mostly suppliers to domestic firms. On the other hand, clothing firms are importing a great deal of their inputs and transforming them for sale mostly in the domestic market, which has already been highlighted when characterizing the textile-clothing sector in Spain.

In conclusion, the analysis of the structural and relational characteristics of the sample confirms its statistical validity. Moreover, any peculiarity found in the sample characteristics validates one of the central arguments of the book: the heterogeneity of the samples with respect to the participation and exploitation of territorial advantages. This aspect is also evident in the relationship between sample characteristics, which has also been discussed in Chapter 4. The nature of the relationships found between the variables that characterize the sample firms and their international activity are summarized in Table 5.14.

5.4 Conclusions

This chapter has established a methodology of analysis that tries to be as objective as possible by following four principles in the geographical delimitation of an industrial district: maximum disaggregation of data, diversification of produc-

Table 5.13 Association between subsector and international activity

	Exports and imports	*Only exports*	*Only imports*	*No exports or imports*
Textile fibres	–2.1	–0.4	0.1	**3.2
Clothing	–0.6	–0.9	*1.9	0.6
Home textiles	**2.5	0.9	–1.1	–3.6
Other activities	–0.2	0.3	–0.8	0.5

Notes
** significant at 5%; * significant at 10%; Chi-square's p-value=0.016.

Table 5.14 Nature of the relationships between sample variables

	Size	*Age*	*District*	*Subsector*
Age	Related	–	–	–
District	Unrelated	Unrelated	–	–
Subsector	Unrelated[1]	Unrelated	Related	–
Internationalization	Related	Unrelated[2]	Unrelated	Related

Exceptions
1 Large firms in the textile fibres subsector.
2 Firms with over 40 years of activity exporting and importing and firms with less than ten years of activity neither exporting nor importing.

tion units, spatial continuity and high dependency. The criteria were applied to identify the largest textile-clothing industrial district in Spain, located in the area of Alcoi-Ontinyent within the Comunidad Valenciana. Moreover, the chapter has described the methodology used in the construction of a questionnaire that was distributed to firms affiliated to the textile-clothing association ATEVAL, of which 201 responses were valid. Finally, in order to confirm the validity of the sample, the characteristics of the sample firms and the relationship between those characteristics were analysed.

From the analysis of those relationships, it was found that firm size and age are related and international activities depend on all characteristics considered (size, age, district location and subsector). When comparing district to non-district firms, there are no significant differences regarding size and age. However, home-textiles firms are over-represented in the district, whereas clothing firms are over-represented outside the district. At the same time, internationalized firms are over-represented in the home-textiles subsector and importers are over-represented in the clothing subsector. For this reason, internationalized firms are over-represented in the district and importers are over-represented outside the district. Nevertheless, it is important to bear in mind that the association between international activity and the district is due to the subsector composition of the sample. This aspect will be further developed in the next chapter.

6 Findings from the Valencia textile-clothing industrial district

The material presented up to this chapter gives rise to a number of hypotheses that will be proposed and tested for the sample described in the previous chapter. The verification of these hypotheses will allow the drawing of conclusions about the influence of the territory and of production specialization on the business activities of traditional manufacturing firms, viewed through the lens of the textile-clothing sector. To that end, the questionnaire presented in the previous chapter was distributed to a sample of textile-clothing firms. The current chapter describes the various analyses carried out and conclusions obtained from testing the hypotheses initially formulated against the information thereby obtained.

The chapter is organized around six subsections. The first is purely descriptive and studies the strategic environment within which the sample firms develop their operations. The second describes the operationalization of the variables present in the initial hypotheses and the preliminary statistical analysis carried out. The remaining subsections test the hypotheses derived from the review carried out previously. The objectives set for the chapter are:

- To show how the perceptions of the respondents can be influenced by the type of location (within an industrial district or outside) and by the production activity developed (either the district's main activity or otherwise). This double analysis of the district effect and the subsector effect (McGahan and Porter, 1997; Chang and Sing, 2000; Hawawini *et al.*, 2003) should provide enough statistical evidence to confirm whether that influence exists or not, in which direction and which aspects are more visible. The study of managerial perceptions is important because, as shown by the literature, these determine strategic decisions (Hambrick and Mason, 1984; Lorhke *et al.*, 2004; Chen *et al.*, 2007).
- To confirm that the district location and the subsector are important determinants of the firms' international activity, both with respect to the decision to internationalize and with respect to performance in international markets, measured in terms of export and import intensity. This aspect is investigated here as the correct management of international markets has been pointed out to be a key response to the challenges brought by globalization (Dicken, 2003; Buckley and Ghauri, 2004; Buxey, 2005).

6.1 The firms' strategic environment

Chapter 4 has traced the recent evolution of the textile-clothing industry across Europe, which shows the uneven impact of globalization on the sector. It is conceivable that differences in national policies, specialization patterns and firm structure all contribute towards explaining the heterogeneous evolution of the sector in recent years. At the same time, the organizational structure of textile-clothing SMEs tends to be centred on their clustering in industrial districts. This organizational structure is an additional source of heterogeneity in the performance of textile-clothing firms. Naturally, there are other sources of heterogeneity in firm performance, such as their strategic environment or own strategy. Moreover, care must be taken to ensure that an apparent heterogeneity is not simply the result of methodological aspects of an investigation.

With these issues in mind, it is necessary, prior to testing the proposed hypotheses, to understand the strategic environment where the textile-clothing firms sampled actually move and develop their activity. This work follows three stages: first, an analysis of scenarios; second, a SWOT analysis; third, an analysis of the strategic needs and public policy demands.

6.1.1 Analysis of scenarios

A scenario can be defined as a description of the present and/or future conditions faced by economic agents, in this case firms (Hitt *et al.*, 2007; Johnson *et al.*, 2007). The scenarios technique is complementary to other methods, such as the SWOT analysis or the analysis of strategic needs, and it has many applications (Kotha and Orne, 1989; Kotha and Swamidass, 2000; Joshi *et al.*, 2003; Kathuria and Porth, 2003). The one that is most relevant to our objectives is the study of the evaluation made by the respondents about two important aspects: the current and future situation of the sector and the evolution of their exports and imports.

Evaluations of the current state of the sector and its future evolution

The opinion of the respondents about the competitive position of the textile-clothing sector is evaluated according to two complementary diagnoses: its current situation and the perceptions of its future evolution. To study the current situation faced by those firms, the following question was included in the questionnaire: 'How would you classify the current state of your firm?'. Three answer options were given: good/average, poor, and bad. To measure the perceptions of the future evolution, the following question was introduced in the questionnaire: 'What do you think will be the future evolution of your firm?'. The answer options given were: improve, stay the same, and worsen. Figure 6.1 offers a graphical summary of the results obtained. Only half of the respondents believe their firms to be in an average or good situation. Their views about the future can be regarded as pessimistic, as 43 per cent think that the situation will worsen.

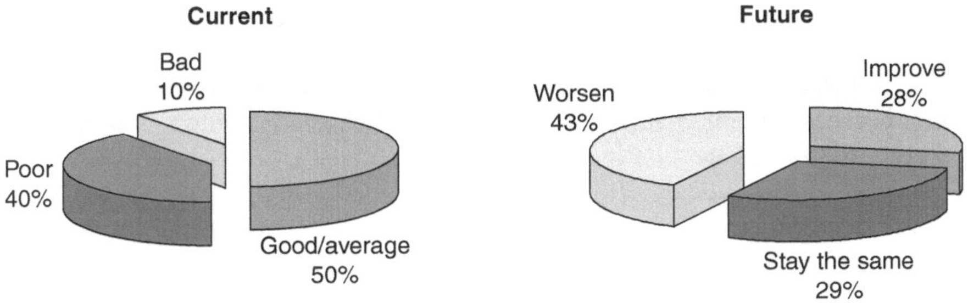

Figure 6.1 Evaluation of the current and future situation.

The main question that arises from this analysis is whether there is any link between the perception of the current situation and the evaluation made of its future evolution. This question is answered using contingency tables, which allow the study of the degree of correlation or independence between the various evaluations (Hair *et al.*, 2000). The resulting analysis indicates that there is a certain degree of association between the answers to the two questions. For example, those that classified the current state of their firm as 'good/average' also had a significantly more optimistic perspective of the future than those classifying it as 'poor/bad' (Table 6.1).[1]

The three main findings up to this point can be summarized as follows: first, the respondents that evaluate the current state of their firm positively are also optimistic about their future evolution; second, among those who evaluate their current state as 'poor', the views about the future are somewhat heterogeneous (they are distributed across all the given options); third, those with a pessimistic opinion about the current state of their firm expect a worsening of the situation. Given these findings, it can be concluded that a large share of the textile-clothing firms are going through a period of poor health and are pessimistic regarding their future. As shown later in this chapter, this pessimistic scenario will influence the firms' strategic decisions.

Evaluations of the evolution of international activity

The other issue debated is related to the situation of activities in international markets (DeMartino *et al.*, 2006). To study this issue, a number of closed

Table 6.1 Relationship between current state and future evolution

	Improve	*Stay the same*	*Worsen*
Good/average	1.5	*1.8	–3.0
Poor	–0.5	–0.9	1.3
Bad	–1.6	–1.6	**3.0

Notes
** significant at 5%; * significant at 10%; Crammer's V=0.198.

questions on two key aspects (exports and imports) were introduced in the questionnaire. There were three answer options: increasing, static, or decreasing. The analysis of exports would allow a judgement on the situation faced by firms selling in foreign markets, while the study of imports would indicate the evolution of purchases in international inputs markets (Lye and Hamilton, 2001).

Figure 6.2 shows the two indicators. From the data it is apparent that, while 53 per cent of respondents say that imports are increasing and 19 per cent say they are decreasing, 23 per cent claim that exports are increasing and 49 per cent that they are decreasing. In both cases, the percentage of respondents stating that their international activity has remained static is about 27 per cent. It is clear that the two indicators are following opposite distributions, with imports increasing strongly and exports decreasing noticeably. This tendency is worrying but was expected from previous discussions (Dicken, 2003; Jones and Hayes, 2004).

Finally, to complete the contextualization of the scenario faced by the firms sampled, it is important to establish whether there is any correlation between the perception of the current and future state of the firm and the evolution of its international activity. This analysis requires four cases: (1) current situation and imports; (2) future evolution and imports; (3) current situation and exports; and (4) future evolution and exports. The results are significant only for case 3 and so only that case is shown in Table 6.2. The corrected residuals show that, in general, when the export level is static or increasing, views are optimistic. On the contrary, when exports are decreasing, the view is that the firm is in a poor or bad state and that it will get worse.

6.1.2 SWOT analysis

The SWOT analysis can be carried out at various organizational levels, from different points of view and using a range of techniques of analysis (Hitt *et al.*,

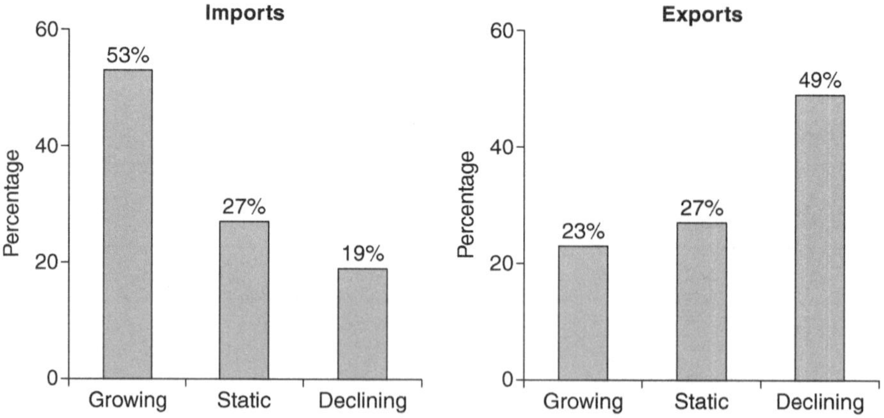

Figure 6.2 Evolution of international activity.

Table 6.2 Relationship between perception of current state and evolution of exports

	Increasing	*Static*	*Decreasing*
Good/average	**2.3	**2.0	–3.7
Poor	–2.0	–1.2	**2.8
Bad	–0.6	–1.5	*1.8

Notes
** significant at 5%; * significant at 10%; Crammer's V=0.236.

2007; Johnson *et al.*, 2007). Similarly to the analysis of scenarios, and in order to understand the strategic environment of the sample firms, the SWOT analysis is carried out jointly for all sample firms (independently of their subsector or of whether they are located inside or outside the district), from the subjective perspective of their managers and representatives, and considering both external and internal factors. The ultimate objective of the SWOT analysis is the construction of a SWOT matrix (Hitt *et al.*, 2007; Johnson *et al.*, 2007). The SWOT matrix presented here reflects all the factors that had previously been pointed out by a panel of experts, and that the respondents perceive as opportunities and threats (competitors, political and economic factors, and so on), as well as the main competitiveness factors (strengths and weaknesses). We follow a two-stage process, first analysing the external environment and then conducting an internal analysis.

For the respondents to evaluate the external environment, they were put before a range of strategic factors along several dimensions (political, territorial, economic, social, cultural and technological) divided into 16 items. The instrument used to measure those perceptions was a five-point Likert scale, where 1 is the most negative value (threat), 3 represents indifference and 5 is the most positive value (opportunity). The quantification was done using the score average. The descriptive statistics generated can be interpreted to answer our starting question.

Table 6.3 summarizes, in ascending order, the score of each factor. According to these results, the respondents have clearly identified as main threats or factors with a very negative impact: *the growing international competition from Asia* (average: 1.34), *lack of political sensitivity* (average: 1.44) and *the importers of finished products* (average: 1.58). Those identified as opportunities or factors with a positive impact were: *the proximity to firms in all stages of the value chain* (average: 3.03), *the EU enlargement* (average: 3.05) and *the threat by the EU of imposing safeguard clauses* (average: 3.19).[2] Moreover, the low standard deviation present in the evaluations (see Appendix) validates the generalization of those results.

The internal analysis to determine the competitiveness factors of textile firms (strengths and weaknesses) was carried out along the same lines as the external environment analysis, with 15 items now being considered, all related to the functional areas of the firms (marketing, finance, etc.). The first step in the analysis consisted of listing the score average obtained for each item using a five-point Likert scale (1 if below the competitors' average, 3 if average and 5 if much above). In this way, the score of each item also measures its contribution towards a competitiveness index. Those factors with average scores below 3 are interpreted as factors whose contribution is negative (weakness), those with an average score equal to 3 are considered indifferent and those with average scores above 3 are regarded as relevant or important.

Table 6.4 shows these scores in a decreasing order. The elements that contribute the most to the competitiveness of these firms are: *the capacity* (flexibility) *to adapt to customer requirements* (average: 3.77), *the technical experience in the sector* (average: 3.75) and *technical and productive flexibility* (average: 3.43). The aspects that contribute the least to competitiveness are: *the presence of their products in international markets* (average: 2.80) and *the development achieved with strategies of cooperation* (average: 2.66). Once again, the low variation of the scores makes the evaluations of the sample firms very precise and reliable.

The results of the SWOT analysis can be summarized in the SWOT matrix (Table 6.5). From a first reading, it is apparent that the main threats and some opportunities (growing international competition and EU enlargement) are similar to those found in other studies of the same geographical area, such as that by Golf *et al.* (2005). The same does not apply to the issue of political sensitivity, which has to be placed in the context in which the questionnaire was delivered, soon after the WTO's liberalization of the textile-clothing trade. With respect to those factors internal to the firm, the main weaknesses relate to the weak strategic development achieved from international activities and cooperation, and the strengths reside in the capacity to adapt to customers' requirements.

Table 6.3 Environmental strategic factors

Item	*Average*
The growing international competition from Asia (China, Pakistan, India, etc.)	1.34
Little political sensitivity to the problems faced by the textile industry	1.44
Importers of finished products	1.58
Large retailers buying directly from emerging markets	1.72
The euro appreciating against the US dollar	1.85
The continuous increase in oil prices	1.88
The progressive disappearance of traditional retailers	2.02
The absence of international harmonization of labels	2.35
The changes in consumption habits for textile products	2.37
The centralization of the purchase function carried out by many distributors	2.42
Most companies using very similar competitive strategies	2.53
The bargaining power of yarn manufacturers	2.62
The dependency on foreign suppliers of machinery	2.84
The proximity to firms in all stages of the value chain	3.03
The EU enlargement	3.05
The threat by the EU to impose safeguard clauses	3.19

Table 6.4 Internal strategic factors

Item	*Average*
Capacity to adapt to customer requirements	3.77
Technical experience in the sector	3.75
Technical and productive flexibility	3.43
Turn-around of samples and designs	3.36
Reputation and brand	3.35
Result of innovative activity (design, packaging, reliability, etc.)	3.25
Decentralized model for decision-making	3.24
Firm size	3.13
Workers' skill level	3.10
Geographical location	3.07
Product prices	3.07
Workers' flexibility	3.06
Establishing and following a strategic plan	3.04
Presence of products in international markets	2.80
Development achieved with strategies of cooperation	2.66

In general, the results obtained in the SWOT analysis allow an understanding of the reasons why many firms feel they currently face adverse conditions. Although there is some heterogeneity in the firms' global evaluation (pessimistic versus optimistic), they offer a homogeneous evaluation of the external and internal determinants of their perceived situation. From the interaction of both aspects, an uneven assessment of strategic needs and policy priorities is to be expected.

6.1.3 Analysis of strategic needs and policy priorities

To be complete, the analysis of the strategic environment of the sample firms requires the identification of which strategies they see as more credible and which public policies are more urgent. With this objective in mind, two lists

Table 6.5 SWOT matrix

Threats The growing international competition from Asia (China, Pakistan, India, etc.) Little political sensitivity to the problems faced by the textile industry Importers of finished products	**Opportunities** The threat by the EU to impose safeguard clauses The EU enlargement The proximity to firms in all stages of the value chain
Weaknesses Presence of products in international markets Development achieved with strategies of cooperation	**Strengths** Capacity to adapt to the customer's requirements Technical experience in the sector Technical and productive flexibility

have been quantified using five-point Likert scales. In the list *credibility of firms' strategies*, 1 stands for 'not necessary', 3 for 'necessary' and 5 for 'very necessary'. In the list *priority of public policies*, 1 means 'not necessary', 3 means 'necessary in the medium term' and 5 means 'urgently necessary'.

Similarly to the population of European textile-clothing firms, our sample firms are very heterogeneous with respect to size, as 23.9 per cent are micro-firms, more than half (55.7 per cent) are small firms and only 3.5 per cent are large firms. For this reason, and taking into account the previous validity results originating in panels of experts, the listed strategies are seen from a corporate point of view, disaggregated in three categories: firm, product and geographical. Within the firm, the respondents have valued aspects related to the sourcing of value chain activities and the ways to implement them. With respect to the product, the main issue has been that of diversification against specialization. Finally, at the geographical level, most issues relate to international markets. As shown in Table 6.6, the respondents do not place a high average value on the proposed options. Despite this, the strategies that are perceived as more important for the textile firms are: *exploring new markets for their products* (average: 3.17), *increasing the product range offered (textiles or otherwise)* (average: 3.08), *specializing in a market niche* (average: 3.04) and *diversifying towards technical textiles* (average: 2.82).

The relatively low urgency of strategic needs perceived by the respondents with respect to many items can be explained by the heterogeneity of the sample. Setting aside this limitation, some interesting links to the SWOT analysis can be detected: (1) the high degree of relevance attributed to internationalization is strongly related to the weakness shown by this aspect; (2) the option of widening the product range may be related to the main strength shown by the firms (flexibility towards customers' requirements); and (3) on the contrary, the

Table 6.6 Strategic needs

Item	*Average*
Exploring new markets for their products (South Asia, Latin America or Eastern Europe)	3.17
Increasing the product range offered (textiles or otherwise)	3.08
Specializing in a market niche	3.04
Diversifying towards new subsectors (technical textiles)	2.82
Higher control of distribution (own shops or franchising)	2.53
Progressively abandon production activities and become a distributor-importer	2.46
Establish strategic alliances with competitors (cooperation)	2.41
Establish agreements for international subcontracting	2.26
Switch to another activity unrelated to textiles	2.15
Multilocalize or reorganize resources internationally (outward FDI)	2.14
Increase the size of the firm	1.83
Reduce the number of customers to better specialize in one or a few of them	1.52
Delocalize the firm to another country	1.52

specialization and diversification towards technical textiles brings together the valuations of other business opportunities that were already apparent in the preliminary stages of consultation with panels of experts.

The strategies considered the least relevant were the ones relative to the *delocation and reduction in the number of customers* (both with average: 1.52) and the *increase in firm size* (average: 1.83). This finding can be explained in several ways. First, delocation is equally associated with a strategy that can only be achieved by the largest firms and is an option that is not well-regarded socially (Dei Ottati, 2002). Second, specialization in a few customers is regarded as a very risky alternative. Finally, a larger size can be viewed as implying a loss of flexibility due to a larger number of employees.

With respect to public policies (Table 6.7), the respondents have in general attributed very high average values to all items (above 3). This means that the priority given to them should be regarded as high. Perhaps the reason behind these high scores is related to the generalized perception of a somewhat passive attitude of policy-makers towards the problems currently faced by the textile-clothing industry. Besides, at a more disaggregated level, a high correlation can be found between these and the results of the SWOT analysis discussed previously. Hence, the two public policies that are seen as more urgent, *pursuing international fairness* (average: 4.59) and *imposing heavier trade barriers* (average: 4.29), seem to respond to the high level of threat perceived in the competition from Asia. The request for support to *access external markets* (average: 4.07) seems to correspond to the interest placed on internationalization. The lower value placed on *facilitating entrepreneurial membership of institutions and associations* (average: 3.03) should be seen in the context of this study, as all those interviewed are members of at least one association (ATEVAL).

From the scenarios perspective, the strategic environment of the textile-clothing firms analysed is full of uncertainty and pessimism, especially among

Table 6.7 Policy priorities

Item	*Average*
Pursue international fairness in the rules of the game for competition	4.59
Imposing heavier trade barriers to limit imports of textile products from outside the EU	4.29
Increase support to improve market access to international markets	4.07
International protection of brand and design	4.01
Improve the image of the national textile in international markets through a quality seal	4.00
Higher financial support for diversification towards new textile products	3.95
Increase support for training of employees	3.74
Increase economic support for lay-offs	3.46
Establish fiscal support to facilitate M&As	3.41
Economic support for entrepreneurial cooperation with related sectors (furniture, lighting, etc.)	3.33
Facilitate entrepreneurial membership of institutions or associations	3.03

those firms that have seen their exports decrease. According to the SWOT analysis, the factors that can explain this situation are related to the threats and weaknesses that have been generally found in traditional manufacturing and that originate both in the globalization of the economy and in internal structural weaknesses (growing competition from Asia, lack of strategic vision, etc.). Among the solutions to this situation, strategic options are valued in a heterogeneous way with respect to their degree of relevance, and policy actions are valued in a homogeneous way with respect to their high level of urgency.

However, the question coming out of this analysis is whether those conclusions can be generalized to all firms in the textile-clothing sector or whether, on the contrary, there are differences related to the territory and manufacturing specialization. This amounts to finding out whether there is any kind of influence or moderating effect on the perceptions previously analysed. It is necessary to first carry out a preliminary statistical analysis that operationalizes the variables on which to apply several statistical tests, and in this way reduce the risks of methodological errors.

6.2 Operationalization and preliminary statistical analysis

The objective of this section is the classification of the variables in order to reduce their complexity and facilitate the interpretation of the results. The first stage was the partition of the sample in distinct but comparable subgroups. Here two criteria have been established: the location and the subsector of activity. These criteria define two subsamples, one classifying firms according to whether they are located (or not) in the Alcoi-Ontinyent industrial district[3] independently of their activity, and another classifying firms according to whether their main textile activity is the production of home textiles (or other textile subsectors), independently of their location. This disaggregation allows the testing of the hypotheses previously derived and the investigation of the district effect and the effect of being present in the main subsector in the district (subsector effect). Ultimately, a broader vision of the influence of both variables on business activity and globalization can be obtained.

In order to achieve those objectives, we proceed by taking the district location and the subsector of each firm as the independent (exogenous) variables, while the valuations obtained with the questionnaire are taken as dependent (endogenous) variables. The operationalization of these variables and some preliminary statistical analyses are presented next.

6.2.1 Operationalization of the independent variables

Industrial district

The questionnaire required the identification of the firm and, using that information and additional data provided by ATEVAL, the location and subsector were identified. After defining the geographical area of the district (Section 4.1), the questionnaires (or firms) were grouped into two types: (1) firms located outside the district, or non-district firms (56.2 per cent); and (2) firms located inside the district, or district firms (43.8 per cent).

Subsector

A preliminary analysis revealed as main activities of these firms: yarn, clothing, home textiles, and other related activities. As the exploratory and descriptive analysis carried out for the sector (Section 4.3) concluded that the territory under analysis is very specialized in home textiles, the firms were divided into two categories: (1) home textiles (45.2 per cent of the sample); and (2) other subsectors (54.8 per cent of the sample).

6.2.2 Operationalization of the dependent variables

One of the main objectives of the chapter is the study of the various types of perceptions (environment, internal, strategies and policies) which aggregate a large number of items (55). The variation observed in those items was studied using five-point Likert scales.[4] The operationalization of the dependent variables was based on the research question and the validation of various techniques of multivariate statistical analysis through the criteria of normality, homoscedasticity and linearity. With respect to the research question, both the degree of heterogeneity of the answers in the groups 'environment' and 'strategies' and the scores obtained by the groups 'internal' and 'public policies' were taken into account in the set of hypotheses used to analyse perceptions. With respect to the techniques of multivariate analysis used to test each of the previous aspects, the conditions of statistical applicability that the data should meet also differ. In particular, the heterogeneity test used is the analysis of variance through the Levene test, which is robust to heteroscedasticity,[5] especially if the sample has a high number of observations and the subsamples have similar or equal size (Stevens, 1996).

Before proceeding, it is necessary to establish an intermediate unit of analysis, called *construct*, between the groups and the items. The possible influence of the district location and subsector on the groups or underlying structures will be tested under matrix form. In turn, the underlying structures are identified using factor analysis.[6] According to Hair *et al.* (2000), factor analysis can be used with two main purposes: (1) to analyse the relationships among a large number of variables and explain these relationships in terms of their common underlying dimensions, also called *factors* (Summary – S); and (2) to condense the information obtained from a number of original variables into a smaller number of variables (factors or constructs) while minimizing the loss of information (Reduction – R). Given that the objective is explicitly to group variables, the factor analysis used will be of type R.[7]

The factors or constructs for each group were extracted using the method of *principal components analysis*[8] which, compared to other extraction methods, such as common factor analysis, allows a justification for the maximum proportion of explained variance represented in the original series of variables. The application of factor analysis and the method of principal components to the items of the environment and strategies groups (16 and 13, respectively) allowed a reduction to nine factors or constructs (Table 6.8).[9]

The interpretation of the new operationalized variables in the environment group is as follows:

- *Rivalry*: this factor represents 18.15 per cent of the variance of the environment group. It includes the items: 'most companies using very similar competitive strategies', 'the threat by the EU to impose safeguard clauses' and 'the centralization of the purchase function carried out by many distributors'.
- *International competition*: this factor represents 11.38 per cent of the variance of the environment group. It includes the items: 'the growing international competition from Asia (China, Pakistan, India, etc.)', 'importers of finished products' and 'large retailers buying directly from emerging markets'.
- *Market*: this factor represents 8.3 per cent of the variance of the environment group. It includes the items: 'the euro appreciating against the US dollar', 'the proximity to firms in all stages of the value chain' and 'the changes in consumption habits for textile products'.
- *EU enlargement*: this factor, which represents only 7.7 per cent of the variance of the environment group, only includes the item 'EU enlargement'.
- *Other factors*: this last factor is the least representative of the environment group (with 7.2 per cent of the variance) and groups the items: 'the absence of international harmonization of labels', 'the progressive disappearance of traditional retailers' and 'little political sensitivity to the problems faced by the textile industry'.

The definition of the operationalized variables in the strategy group is as follows:

- *Production*: this first factor in the strategy group represents a variance of 27.56 per cent of the variance of the strategy group and includes the items: 'delocate my firm to another country', 'switch to another activity unrelated

Table 6.8 Summary of factor analysis

Statistics	*Environment group*	*Strategy group*
KMO	0.677	0.778
Sig. Bartlett	0.000	0.000
Determinant	0.100	0.075
Explained variance	52.93%	56.30%
Factors	1 Rivalry	1 Production
	2 International competition	2 Exploring markets
	3 Market	3 Specialization and distribution
	4 EU enlargement	4 Growth
	5 Other factors	

to textiles', 'multilocate or reorganize resources internationally (outward FDI)' and 'establish agreements for international subcontracting'.

- *Exploring markets*: this factor represents 11.51 per cent of the variance of the strategy group and is formed by the following items: 'exploring new markets for their products (South Asia, Latin America or Eastern Europe)', 'diversifying towards new subsectors (technical textiles)' and 'increasing the product range offered (textiles or otherwise)'.
- *Specialization and distribution*: this factor represents 8.82 per cent of the variance of the strategy group. It contains the items: 'higher control of distribution (own shops or franchising)', 'reduce the number of customers to better specialize in one or a few of them', 'progressively abandon production activities and become a distributor-importer' and 'specialize in a market niche'.
- *Growth*: this last factor in the strategy group represents 8.39 per cent of the variance. It includes the items: 'increase the size of the firm' and 'establish strategic alliances with competitors (cooperation)'.

The purpose of the first set of hypotheses is to explain the differences found in the scores of the groups 'internal' and 'public policies' for each subsample defined by the district location and the subsector. However, we face important limitations imposed by ordinal scales, and certain conditions of applicability (normality) are not met. Hence, parametric techniques such as ANOVA, discriminator analysis or linear regression models cannot be used.[10] Instead, we apply an ordinal regression model to each item separately.[11] Finally, the hypothesis relative to international activity is tested using two of the most common indicators: the start of the international activity and its intensity.

- *Start of international activity*. The questionnaire included a question about the year when the firm had started exporting or importing. To operationalize the variable, this year had to be compared to the year of registration of the firm, according to the records of ATEVAL. The result of this operation was the number of years each firm took to start importing or exporting.
- *Intensity of international activity*. The questionnaire also asked about the average share of imports (exports) with respect to purchase (sales).

6.2.3 Preliminary statistical analysis of the internationalization variables

Before hypothesis testing, the conditions of applicability of multivariate analysis to the internationalization variables were investigated (Hair *et al.*, 2000). With this purpose, we first tested for normality and then for homoscedasticity. The normality test was conducted using Q-Q diagrams and the Kolmogorov–Smirnov test.[12] This analysis revealed that the data for the four operationalized variables do not meet the normality conditions. To correct for non-normality and to be able to apply parametric tests, the data had to be transformed according to the trial-and-error procedure outlined in Hair *et al.* (2000). Given that there was no single procedure for transforming the four variables, each one was corrected differently, as shown in Table 6.9.

In the following stages we verified whether the transformations carried out may have had the desired effect. Inspection of the diagrams (Figure 6.3) together with the result of the Kolmogorov–Smirnov test (Table 6.9) for the transformed variables allow the conclusion that the transformed data follow a normal distribution.

6.3 The effect of the industrial district location on perceptions

Industrial districts have been described as a specific type of territorial network. Considering the specific social characteristics of the territorial networks that intensify interpersonal relationships, the organization of production that allows specialization and interdependence between firms, and the economic and social impact of that business environment, the following empirically testable hypotheses can be formulated:

> *Hypothesis 1*: District firms have more homogeneous and more positive managerial perceptions of the environment and competitive position than non-district firms.
>
> *Hypothesis 2*: District firms have more heterogeneous and more negative managerial perceptions of strategic needs and policy priorities.

The statistical analysis presented in the previous section allows a formal testing of the hypotheses laid out above, linked to the possible influence of district location on managerial perceptions. The SWOT analysis has been carried out, distinguishing external and internal aspects, but the effects of a district location can also be seen with respect to a higher homogeneity in the environmental perceptions and a higher valuation of internal strengths. Therefore, hypothesis 1 should be split into two (1a and 1b). In our analysis the perception of strategic needs and policy demands has been analysed in two separate groups and, in turn, the influence of the district location exists in both the heterogeneity of strategic needs and the demand for public policies. Hence, hypothesis 2 should also be disaggregated into two (2a and 2b) (Figure 6.4).

Table 6.9 Results of the transformation of the internationalization variables

Variable	*Transformation*	*P-value K-S*
Number of years to start exporting	Log (variable)	0.055
Number of years to start importing	Squared root (variable)	0.200
% sales exported	Log (–log((100 – variable)/100))	0.077
% purchases imported	Log (variable/(100 – variable))	0.067

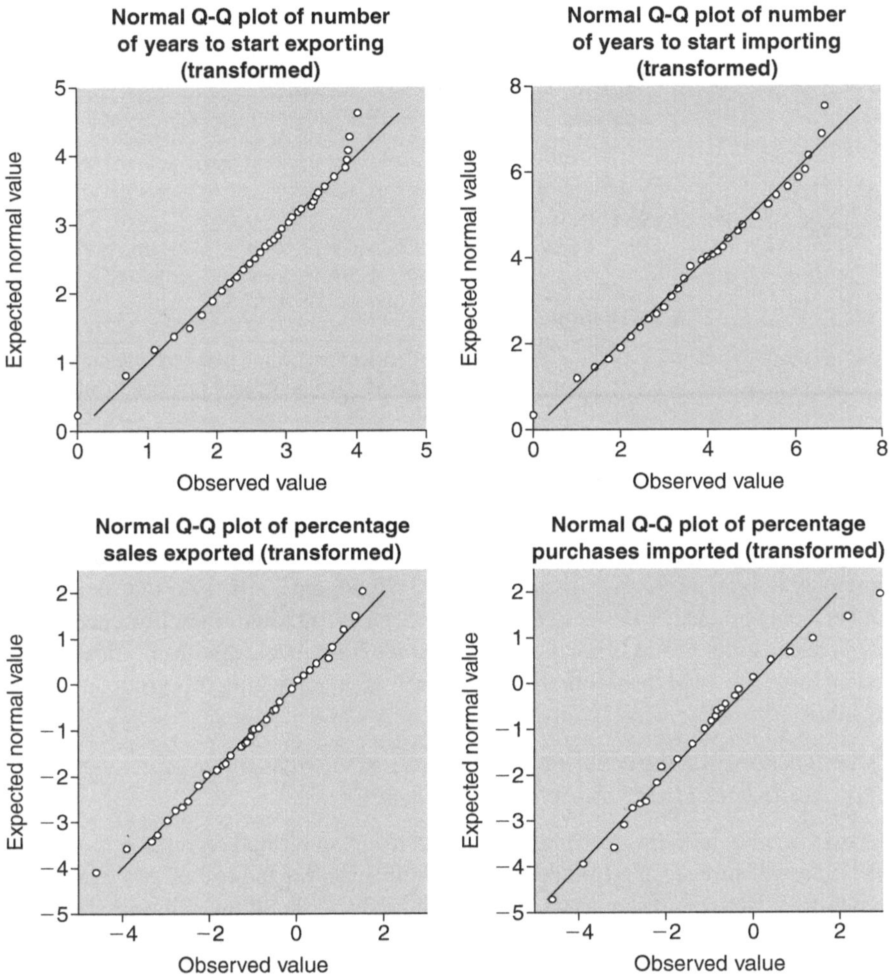

Figure 6.3 Normality test on the transformed internationalization variables.

6.3.1 The district effect on the perceptions of the SWOT analysis

District firms have a more homogeneous evaluation of the environment compared to those located outside the district (hypothesis 1a)

The results obtained with the statistical analysis (Table 6.10) indicate that in the *environment* group the variance of evaluations 0.102 for non-district firms and 0.092 for district firms; this means, in line with the hypothesis advanced previously, that the second group has a more homogeneous perception of the environment than the first. However, such difference is not statistically significant (Levene

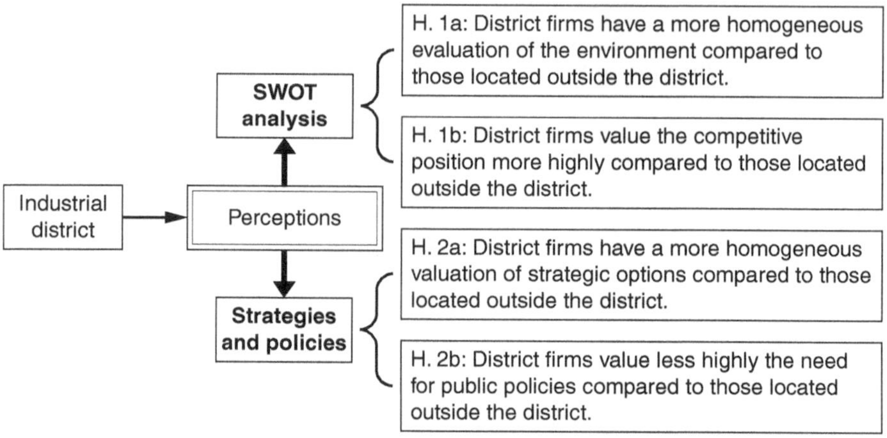

Figure 6.4 Diagrammatic representation of the district effect hypotheses on perceptions.

test *p*-value: 0.535).[13] With respect to the scores of the five factors or constructs previously obtained through factor analysis, the conclusion is also that there is a higher homogeneity in the answers of district firms, although the difference compared to the other group is not significant. At the item level, significant differences in variances have not been found for any of the 16 items within this group.

District firms value the competitive position more highly compared to those located outside the district (hypothesis 1b)

As explained before, this hypothesis is tested using an ordinal regression model, in particular the ordered logit model.[14] The justification for the use of this technique is two-fold: first, the dependent variable or item does not follow a normal distribution, which prevents the use of OLS regression models; second, it is due to the ordinal character of the answers. With this technique, which generalizes the logit regression for ordinal variables with multiple answers, we intend to determine how the probability of occurrence of each state of the dependent variable is conditioned upon the states of the independent variables (district location and subsector).[15] Ultimately, by studying the significance and sign of the estimated coefficients (β), we can obtain solid evidence on the influence of the independent variables and on the probability of occurrence of the lowest scores for each item.[16]

The results obtained with the ordinal regression (Table 6.11) indicate that in the internal group the district location is significant only for two items: *reputaci* and *technexp*. Given that the coding used is district firm = 1 and non-district firm = 0, and the coefficients of the *reputaci* and *technexp* variables are negative, the results can be interpreted as meaning that the lowest scores have been found for the district firms.[17] Hence, contrary to expectations, the lowest scores have a higher probability to be found in district firms than in non-district firms.

Table 6.10 Evaluation of the environment with respect to location

Group/Construct	*Non-district firms*	*District firms*	*LEVENE*	*Non-district firms : District firms*
Environment	0.102	0.092	0.535	82/109
Rivalry	1.214	0.849	0.242	72/95
International competition	1.135	0.907	0.345	72/95
Market	0.995	1.008	0.791	72/95
EU enlargement	1.052	0.970	0.975	72/95
Other factors	0.951	1.022	0.651	72/95

Note
The last column indicates the non-district firm/district firm ratio of valid cases. When the sum of the numerator and denominator is lower than 201 the missing answer to an item has been regarded as invalid by the statistical technique used in each case.

As a corollary of this analysis we can state that hypothesis 1 does not seem to be verified. That is, the district location does not significantly change the strategic perceptions. Hence, the views expressed by the managers of the textile-clothing sector do not significantly differ according to whether they are located in an industrial district or not. As a consequence, the inter-organizational relationships with a territorial base do not seem to materialize in more homogeneous opinions, nor in higher valuations of their competitive position. This last result can be interpreted in different ways: (1) the difficulties of the textile-clothing industry are visible in any territory; (2) the firms have similar strategic capabilities (flexibility); (3) there is not enough cooperation between district firms (interconnection and shared resources); and (4) as a consequence of the previous interpretations, the similarity of perceptions can be seen as a sign of erosion of the role of the territory (Amin, 1993; Cairncross, 1997).

6.3.2 The district effect on the perceptions of strategies and policies

District firms have a more heterogeneous valuation of strategic options compared to those located outside the district (hypothesis 2a)

In this case, the test statistic (Table 6.12) indicates that in the strategies group the non-district firms' variance (0.320) is, as expected, slightly lower than the district firms' variance (0.415). However, the Levene test (0.531) reveals that

Table 6.11 Regression results for the internal group

Variable	*District*	*Subsector*	*G. fit*	*Number of cases*
TECFLEX	0.173	***0.235**	467.78	196
INNOVA	0.973	****0.443**	420.27	194
SIZE	0.985	****0.054**	521.58	197
LOCATE	0.195	****–0.125**	443.41	197
DECENTR	0.329	****0.238**	376.89	193
WORKFLEX	0.576	***0.018**	541.96	197
SKILL	0.660	****0.596**	451.03	197
REPUTACI	****–0.844**	****–0.011**	500.00	197
PRICE	0.144	0.123	433.83	197
RENEW	0.880	*****0.818**	470.60	192
STRPLAN	0.437	****0.079**	444.44	195
REQUIREM	0.236	****0.444**	452.24	198
TECHNEXP	***–0.475**	***–0.016**	431.89	198
PRESENIT	0.164	****0.147**	523.11	190
DEVCOOP	0.305	****0.472**	465.99	186

Notes
The significant coefficients β are given in bold. * Significant at 10%; ** significant at 5%; *** significant at 1%. The goodness of fit or deviance is a concept that can be found in many statistics manuals (for example, Gujarati, 2003). This statistic, which can take any positive value, is the difference between the log-likelihood of the observed values and the adjusted values. Hence, the lower the deviance the better the model is.

such a difference is not statistically significant. At the construct level, we can state that opinions are homogeneous within both non-district firms and district firms, the exception being *exploring markets*. In particular, as expected, the district firms (variance: 1.214) are less in agreement regarding the degree of necessity of the strategies grouped under that heading than the non-district firms (variance: 0.773). This result shows that inside the industrial district some firms see the adoption of those strategies as very relevant, whereas others see them as being of little relevance.

The internal heterogeneity of the district model shown in the data is in line with previous research (for example, Camisón, 2004; Molina and Martinez, 2004; Giuliani, 2007). This diversity is significantly higher with respect to strategic options of multilocation, entering new international markets or international sourcing. In addition, at a more disaggregated level, the higher disparity of opinions is also observed in the value placed on the item '*widening the range of textile products*'. This result can be interpreted as meaning that, while some district firms see it as very important to increase the variety of their supply range, others think in the opposite way.

District firms value less highly the need for public policies compared to those located outside the district (hypothesis 2b)

The results of the logit analysis (Table 6.13) show that in the group of public policies, the district effect on managerial perceptions is significant only for three items: *associat*, *image* and *suptrain*. This result can be interpreted as meaning that the lowest values are found in the district firms and so show a higher probability of occurring in those firms than in the others. For that reason, the urgency of the policy demands of district firms, compared to non-district firms, seems lower, especially with respect to the promotion of associations, improvement of the country image abroad and increased funding for training.

The results obtained allow the acceptance of hypothesis 2 with some caveats. There is some statistical evidence confirming, although weakly, the internal heterogeneity of the district in the construct '*exploring new international markets*'

Table 6.12 Evaluation of strategies according to location

Group/Construct/-items	*Non-district firms*	*District firms*	*LEVENE*	*Number of cases*
Strategies	0.320	0.415	0.531	88/112
Abandon production	1.136	0.902	0.210	78/101
Explore markets	0.773	1.214	*0.073	78/101
Specialization and distribution	0.862	1.063	0.778	78/101
Size	1.034	0.968	0.760	78/101
• Widen range of textile products	0.977	1.620	*0.050	78/101

Note
* significant at 10%

Table 6.13 Regression results for the policies group

Variable	*District*	*Subsector*	*G. fit*	*N. cases*
ASSOCIAT	***–0.433**	****0.008**	608.19	194
SUPLAYOFF	0.984	***0.258**	605.25	196
PROTECIT	0.744	****–0.002**	521.13	195
ACCESSIT	0.953	****0.084**	506.49	195
FAIRGAME	0.409	*****–0.343**	303.38	196
IMAGE	***–0.476**	*****0.403**	510.21	196
SUPFIS	0.848	*****–0.123**	597.67	193
SUPDIVER	0.243	*****–0.274**	513.53	195
SUPCOOP	0.686	****0.135**	590.27	194
TRADEBAR	0.126	*****–0.066**	421.47	198
SUPTRAIN	****–0.602**	****0.081**	550.07	196

Notes
The significant coefficients β are given in bold.
*** significant at 1%; ** significant at 5%; * significant at 10%.

and in the item '*widening the range of textiles products*' (hypothesis 2a). It reduces the policy demands in 3 of the 11 items analysed (hypothesis 2b).

Having come to this point, we can now provide an answer to the question of the influence of the territory on managerial perceptions. Table 6.14 presents a summary of the results obtained for the hypotheses about the effect of the district location on perceptions. From this it can be concluded that the district location seems not to have any influence on the perceptions of the SWOT analysis and has a relative influence on the perceptions of strategic needs and policy demands. Therefore, the strategic options and public policies pointed out by Abernathy *et al.* (2006), Zucchella (2006) or Zourek (2007) as solutions to the current crisis in the textile-clothing sector should take into account both the district location and the firm-specific strategy. This recommendation is based on the general consensus according to which subjective managerial perceptions condition strategic decisions (Hambrick and Mason, 1984; Lorhke *et al.*, 2004), and the fact that and in times of difficulty turnaround strategies are required (Filatotchev and Toms, 2003; Robbins and Pearce, 1992). Hence, the shared pessimistic managerial perceptions, as well as the visible signs of the erosion of the competitiveness of the industrial district, should be taken into account in future research and in improving the efficacy of future strategic and policy solutions.

6.4 The district effect and international activity

As shown in Chapter 3, the degree of internationalization of the network to which the firm belongs seems to have an impact on the firm's own activity, such that it can help explain the exporting start and intensity (Johanson and Mattson, 1988). However, there are few studies that look at whether a similar impact can be found on the importing side or whether the district location and the subsector play any moderating role with respect to imports (Lye and Hamilton, 2001).

Table 6.14 Summary of hypothesis testing results on district location and perceptions

Hypothesis 1: Locating in an industrial district makes more homogeneous and improves the perception of the SWOT analysis compared to a non-district location.	
Hypothesis 1a: District firms have a more homogeneous evaluation of the environment compared to those located outside the district.	*Not verified*
Hypothesis 1b: District firms value the competitive position more highly compared to those located outside the district.	*Not verified*
Hypothesis 2: Locating in an industrial district makes more heterogeneous and deteriorates the perception of strategic and policy needs compared to a non-district location.	
Hypothesis 2a: District firms have a more heterogeneous valuation of strategic options compared to those located outside the district.	*Partly verified*
Hypothesis 2b: District firms value less highly the need for public policies compared to those located outside the district.	*Partly verified*

Hence, the objective that will be pursued in this section is to analyse, from the twin perspective of exports and imports, the influence of the district location on the start and intensity of internationalization.

In this way, internationalization (start and intensity) is taken as a variable dependent on firm location. As before, the variable 'start' was operationalized as the start of importing activity and start of exporting activity, while for the variable 'intensity' the percentage of imports in purchases and the percentage of exports in sales were defined. The four measures were transformed in order to meet the normality criterion (Section 6.2.3). This statistical analysis allows a parametric testing of the hypotheses developed previously, which established the possible influence of location on the start and intensity of internationalization:

> *Hypothesis 3*: The start of international activity will occur earlier for district firms than for non-district firms.

> *Hypothesis 4*: The intensity of international activity will be different in the district and outside.

The analysis of the start and intensity of internationalization distinguishes two types of international activities: imports and exports. Hence, the two more general hypotheses need to be split into four more specific hypotheses (hypotheses 3a, 3b, 4a and 4b in Figure 6.5).

6.4.1 The district effect on the start of internationalization

The dependent variables related to the start of international activity meet the applicability conditions as previously shown (Section 6.2.3). Under these conditions, hypotheses 3a and 3b can be tested parametrically. The results of the

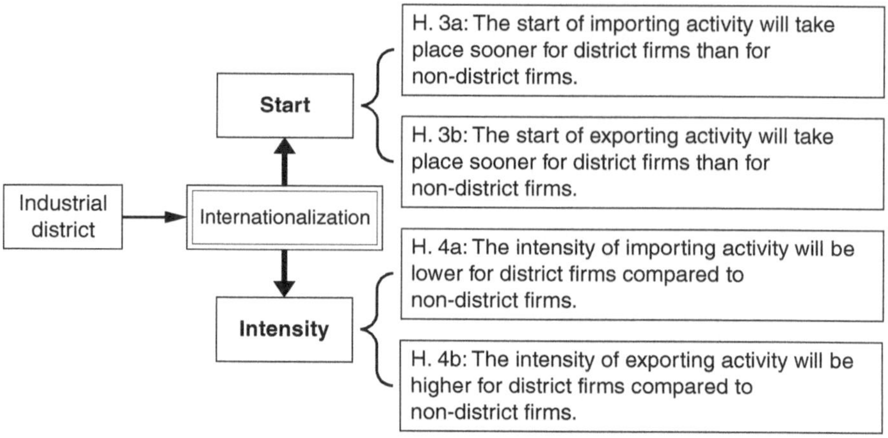

Figure 6.5 Diagrammatic representation of the hypotheses for the district effect on internationalization.

parametric tests carried out for these hypotheses are summarized in Table 6.15. As shown, independently of location, in all cases firms take longer to start importing than exporting. In particular, firms located in the district start importing within less than ten years (average: 9.77) of starting up, while the ones outside the district start importing, on average, 17 years after their creation (average: 17.05). The same tendency is found for exports, as district firms take a little over eight years (average: 8.29) to start exporting, while the other firms need, on average, 14.5 years.

It is now necessary to understand the dispersion of those scores for the transformed data within the various subgroups and to study the statistical significance of the results. Similarly to Hair *et al.* (2000), the first task is carried out using a box plot[18] and the Levene homoscedasticity test, whereas the second task is carried out using the analysis of variance (ANOVA). In both situations,

Table 6.15 Summary of the district effect on the start of internationalization

	Averages		*Statistics*		*Number of cases (Non-district firm/ district firm)*
	Non-district firm	*District firm*	*F-test*	*Levene*	
Years it takes to start importing	17.05	9.77	**0.007	0.243	40/48
Years it takes to start exporting	14.51	8.29	**0.016	0.575	54/62

Notes
** significant at 5%; * significant at 10%.

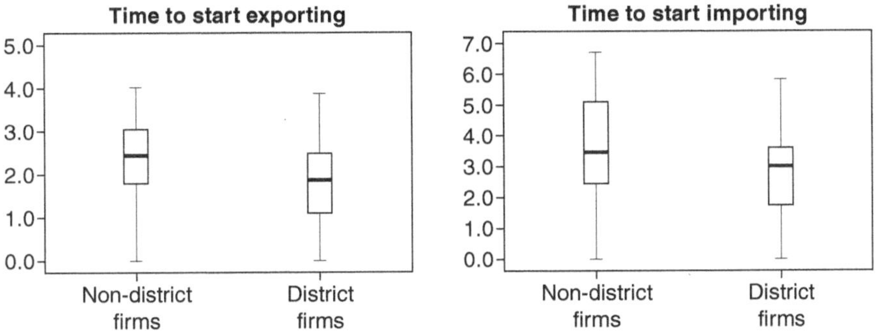

Figure 6.6 Box plot for the start of international activity according to location.

described in Figure 6.6, there are values that indicate the possibility of finding significant differences in terms of centrality, symmetry and variability of the observations. However, it is apparent that dispersion is not very different across cases. This observation takes us back to thinking about the validity of the data. Besides, the statistical results from the Levene test in Table 6.15 show that the data meet the applicability conditions (Levene > 0.05 in all cases).

The significance of the F-test (Table 6.15) indicates that there are significant differences in the average scores achieved in testing the district effect. Therefore, we can accept the third hypothesis. Similarly to Maccarini *et al.* (2003a, 2003b), this finding leads us to conclude that there seems to be evidence supporting the existence of some impact of the district location on the time taken to start international activity; however, unlike previous work, the effect is found here both on the import and on the export sides.

6.4.2 The district effect on the intensity of internationalization

The dependent variables representing the intensity of internationalization (propensity to import and export) meet the conditions of applicability as tested previously (Section 6.2.3), which allows us to test the related hypotheses parametrically. Similarly to the previous section, the intensity of both imports and exports was considered. The most general hypothesis (the impact on the intensity of internationalization) should be disaggregated into two more specific hypotheses (4a and 4b). The data obtained from the parametric tests carried out for the individual testing of these hypotheses are summarized in Table 6.16. It can be seen that the average intensity of international activity is as expected: on average, district firms buy less and sell more in foreign markets than non-district firms. Moreover, the influence of the district location seems to be higher for exporting than for importing.

The dispersion of transformed data (Figure 6.7) indicates that none of the two figures contains very different values in terms of centrality, symmetry or

Table 6.16 Summary of the district effect on the intensity of internationalization

	Averages		*Statistics*		*Number of cases (non-district firm/ district firm)*
	Non-district firm	*District firm*	*F-test*	*Levene*	
% imports on purchases	26.55	23.77	0.838	0.134	40/47
% exports on sales	31.58	38.51	*0.068	0.130	57/75

Notes
** significant at 5%; * significant at 10%.

variability, again confirming the validity of the transformed data. With respect to the significance of the F-test, hypothesis 4b seems to be verified (p-value: 0.068). On the contrary, hypothesis 4a is rejected (p-value: 0.838). Hence, we find a positive impact of the district location on exporting intensity.

To what extent does the industrial district location influence import and export activities? Table 6.17 summarizes the test results for the hypotheses related to the district effect on international activity. First, the location of a firm in the district seems to be an important factor influencing the activity related to international export markets. Second, that influence varies according to the actions related to international import markets. Finally, the results are in line with previous results in the literature on manufacturing firms, such as those obtained by Costa and Viladecans (1999) for Spanish firms or Becchetti and Rossi (2000a, 2000b) for the case of Italy.

These results highlight the diversity of responses of small and medium textile-clothing firms facing globalization, with some firms being more focused on local markets and other firms being regularly present in international markets (Julien, 1996). The latter is the most common type of response in industrial districts. So, similarly to Johanson and Mattson (1988), Bell (2005) or Belussi and Sedita (2009), the data seem to confirm the existence of a growing link between the

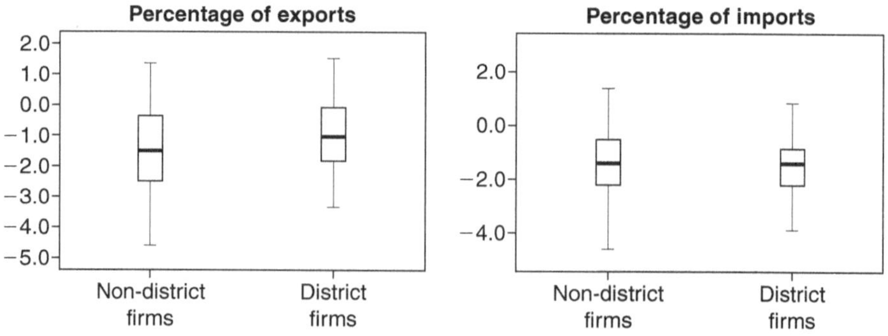

Figure 6.7 Box plot of the intensity of international activity according to location.

Table 6.17 Summary of hypothesis testing results on district location and internationalization

Hypothesis 3: The start of international activity will take place sooner for district firms than for non-district firms.	
Hypothesis 3a: The start of importing activity will take place sooner for district firms than for non-district firms.	*Accepted*
Hypothesis 3b: The start of exporting activity will take place sooner for district firms than for non-district firms.	*Accepted*
Hypothesis 4: The intensity of international activity will be different inside the district and outside the district.	
Hypothesis 4a: The intensity of importing activity will be lower for district firms compared to non-district firms.	*Rejected*
Hypothesis 4b: The intensity of exporting activity will be higher for district firms compared to non-district firms.	*Accepted*

firms and the relational model to which they belong. According to this, it is to be expected that the social and organizational characteristics found in the industrial district inhibit in a way the relocation tendency that is currently common in the manufacturing sector.

In conclusion, if the internationalization strategy is closely related to its stability, especially in mature industries, the more global orientation of industrial districts can serve to preserve its survival if, as suggested by Mariotti (2004) or Zucchella (2006), the local context is renewed and contributes to the competitiveness of the district. From another point of view, that greater commitment to international markets should be taken as an organizational reference for non-district firms, as it helps explain, in part, its internal heterogeneity. These last two aspects suggest that some solutions brought forward to overcome the difficult situation in the textile-clothing sector should consider whether firms are located in the district or not.

6.5 The subsector effect on perceptions

In the context of traditional manufacturing, there is little empirical evidence on the impact of belonging to the main subsector of a territory on firm performance. However, the textile-clothing industry, like the rest of traditional manufacturing, is to a large extent composed of geographically concentrated SMEs (Goglio, 2002). Each concentration of firms specializes in a given subsector in which it develops a set of human and technical capabilities that make the firms in that subsector more competitive. The benefit of locating within the concentrated territory may be less for those firms that dedicate themselves to related subsectors with respect to the main subsector, possibly because the competition–cooperation relationships between firms operating in the same subsector will be more frequent and more intense than those taking place outside the same subsector. As a result, the sharing of resources and knowledge is more intensive in the main subsector than in related subsectors. The interaction between territory and specialization will determine firm performance, as well as the type and homogeneity of managerial perceptions of the strategic environment, needs and policies:

> *Hypothesis 5*: Firms operating in the main subsector of the territory have more positive and more homogeneous perceptions of business performance.

> *Hypothesis 6*: Firms operating in the main subsector of the territory have more negative and more heterogeneous perceptions of strategic needs and policy priorities.

The operationalization, investigation of the conditions of applicability, and transformations already carried out will allow the parametric testing of the possible influence of the subsector on managerial perceptions. With this analysis we intend to, at least partly, shed some light on an issue that impacts on strategic decisions relative to the management of operations in very specialized territories but that, until now, has not been studied at length (Nassimbeni, 2003). This study is very important for the performance and competitiveness of the firm (Skinner, 1969; Hayes and Wheelwright, 1984; Devaraj *et al.*, 2004). This is especially true when globalization puts pressure on the decisions of what and where to produce (Buckley and Ghauri, 2004; Zucchella, 2006).

In the strategic analysis carried out previously, both external and internal factors were taken into account. The theory allows us to distinguish two avenues of influence: (1) more homogeneous perceptions about the environment; and (2) higher score for internal strengths. Hence, each of the previous hypotheses should be split into two (5a, 5b, 6a and 6b). In this way, as summarized in Figure 6.8, we arrive at a total of four hypotheses.

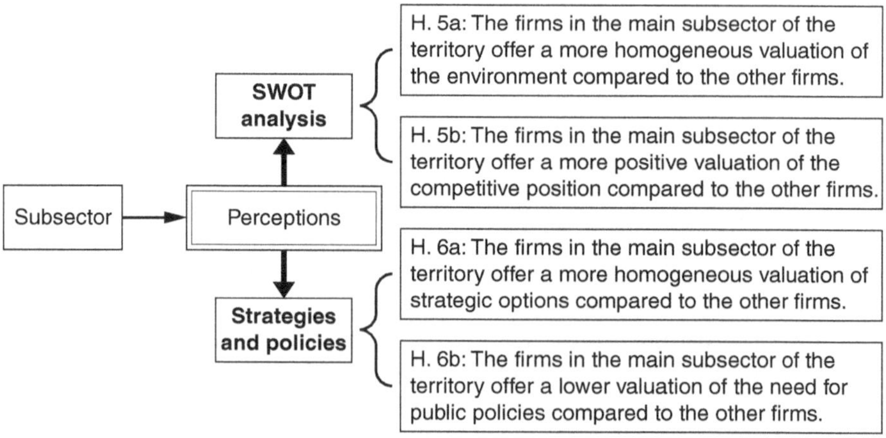

Figure 6.8 Diagrammatic representation of the subsector effect on perceptions.

6.5.1 The subsector effect on the perceptions of the strategic analysis

The firms in the main subsector of the territory offer a more homogeneous valuation of the environment compared to the other firms (hypothesis 5a)

The results obtained from the analysis of homogeneity of variances (Table 6.18) indicate that, at the group level, the variances do not present significant differences; at the construct level, the variability of the 'other factors' construct is significantly different; and at the item level, there are significant differences in five of the 16 items, with four of them, similarly to the 'other factors' construct, having the expected sign (lower dispersion).

The construct 'other factors' includes the following items: 'progressive disappearance of traditional retailers', 'absence of an international harmonization of labels' and 'little political sensitivity with respect to the problems of the textile sector'. The items with a significantly more homogeneous variance are linked to policy issues, such as 'the threat of using safeguard clauses', or to some sources of competitive rivalry ('growing international competition' or 'bargaining power of yarn manufacturers'). In conclusion, the results obtained seem to partly verify hypothesis 5a. This implies recognition that, independently of the average value of the scores (Table A.2.6), the main subsector in the territory (home textiles) shows more homogeneous valuations.

The firms in the main subsector of the territory offer a more positive valuation of the competitive position compared to the other firms (hypothesis 5b)

The logit analysis carried out previously (Table 6.13) shows a generalized influence of the subsector except in the item *price*. All items except *locate* have the expected sign. Hence, if in 13 of the 15 variables the statistical significance exists and the interpretation derived from the positive sign of the coefficients is as expected, we can conclude that the productive specialization of the firms in the main subsector in the territory has a positive influence on the valuation of their competitive position.

Nevertheless, the negative sign on the significance of the item *locate* (geographical location of the firm) is not as expected. It would be expected that those firms in the home-textiles subsector, most of them located in the industrial district, would recognize this aspect as a source of competitive advantage. However, the observed valuations indicate otherwise. Two arguments can be made here. The first is that if most firms are located in the same territory they do not perceive an advantage in the comparison as all have access to the same location advantages. Putting it in another way, it could be said that the highest score for location is given, not by the main subsector in the territory (home textiles), but by firms in peripheral subsectors. This result could be due to the strong dependency that firms in peripheral subsectors have relative to firms in

Table 6.18 Valuation of the environment according to the subsector

Group/Construct/-items	*Others*	*Home*	*Levene*	*Number of cases*
Environment	0.093	0.100	0.498	106/84
Rivalry	0.861	1.172	0.593	92/74
International competition	1.107	0.894	0.291	92/74
Market	1.104	0.893	0.372	92/74
EU enlargement	0.965	1.045	0.275	92/74
Other factors	1.189	0.766	**0.021	92/74
• Large retailers buying in emerging markets	0.618	1.124	**0.021	92/74
• Growing international competition	0.377	0.272	**0.037	92/74
• Bargaining power of textile manufacturers	0.581	0.446	*0.075	92/74
• No international harmonization of labels	0.702	0.372	***0.000	92/74
• Threat of using safeguard clauses	1.562	0.961	**0.037	92/74

Notes
*** significant at 1%; ** significant at 5%; * significant at 10%.

the main subsector. The second argument to be made is that, as has already been shown, this result could be interpreted as meaning that the role of the territory is being lost.

In conclusion, hypothesis 5 seems to be validated. This allows us to state that belonging to the home-textiles subsector seems to have an influence over managerial perceptions. Therefore, the vision of crisis that the managers of the textile-clothing sector seem to have varies according to the subsector in which they operate. From this analysis, it can be seen that the productive specialization in the main subsector of the territory materializes in more homogeneous views about the environment and in better views about the firm's competitive position. This conclusion validates an important aspect highlighted in this industry regarding the uneven impact suffered with the changes in the competitive environment (Jones and Hayes, 2004; Golf *et al.*, 2005; Buxey, 2005).

6.5.2 The subsector effect on the perceptions of strategies and policies

To test the influence of manufacturing activity on managerial perceptions of strategic needs and policy demands, two different hypotheses have been established (6a and 6b).

The firms in the main subsector of the territory offer a more homogeneous valuation of strategic options compared to the other firms (hypothesis 6a)

Table 6.19 shows that, according to the prior hypothesis, the valuations made by firms in the home-textile subsector are significantly more homogeneous (variance: 0.284) than those in the subsample for the other subsectors (variance: 0.458). In a way, we can also arrive at the same conclusion when we refer to the constructs 'exploring international markets' (variance home textile: 0.853 versus variance others: 1.152) and 'size' (0.844 versus 1.137). At the item level, the results are along the same lines. As shown in Table 6.19, the variances of the

Table 6.19 Valuation of strategies according to the subsector

Group/Construct/-items	*Others*	*Home*	*Levene*
Strategies	0.458	0.284	****0.012**
Quit production	1.056	0.970	0.387
Exploring international markets	1.152	0.853	*0.091
Specialization and distribution	1.011	1.015	0.322
Size	1.137	0.844	*0.092
• Alliances with competitors	1.650	1.243	**0.020
• Quit production	1.864	1.319	*0.055
• Widen product range	1.497	1.187	*0.074
• International subcontracting	1.475	1.104	**0.037

Notes
*** significant at 1%; ** significant at 5%; * significant at 10%.

items 'alliances with competitors', 'quit production', 'widen product range' and 'international subcontracting' have a significantly lower variance for home-textile firms than for other subsectors.

In conclusion, given that the behaviour expected from the firms in the home textile subsector is in general more homogeneous, it seems reasonable to accept hypothesis 6a.

The firms in the main subsector of the territory offer a lower valuation of the need for public policies compared to the other firms (hypothesis 6b)

The results obtained with the logit analysis, together with the significance of each variable tested (Table 6.15), indicate a visible influence of the subsector on the perceptions of the items in the public policy group. Nevertheless, the (negative) sign of the coefficients is expected only in five of the 11 items: *protecit*, *fairgame*, *supdiver*, *supfis* and *tradebar*. This result supposes, in line with the suggested hypothesis, that the low scores observed and expected for those variables or items are more frequent in the main subsector in the territory (home textiles). Therefore, the effect is as expected for those aspects: a lower valuation of the need for public policies compared to the other subsectors. Unlike the district effect on perceptions, we find six items where the subsector effect is significant, but with a sign opposite to what was expected. This finding indicates a greater need for public policies on these issues. So, hypothesis 6b does seem to be verified.

For this reason, the results obtained from testing the more general hypothesis 6 do not allow us to reject it. Hypothesis 6a can be accepted and the significance of the influence of the subsector is important in all items analysed, although in some cases with a coefficient sign different from the expected. Table 6.20 presents the main test results obtained for the subsector effect on managerial perceptions. It can be seen that the main activity in the territory (home-textile

Table 6.20 Summary of results of hypothesis testing on subsector and perceptions

Hypothesis 5: Belonging to the main subsector in the territory makes more homogeneous and improves the perception of the strategic analysis compared to other subsectors.	
Hypothesis 5a: The firms in the main subsector of the territory offer a more homogeneous valuation of the environment compared to the other firms.	*Partly*
Hypothesis 5b: The firms in the main subsector of the territory offer a more positive valuation of their competitive position compared to the other firms.	*Accepted*
Hypothesis 6: Belonging to the main subsector in the territory makes more homogeneous and worsens the perception of the strategic and policy needs compared to other subsectors.	
Hypothesis 6a: The firms in the main subsector of the territory offer a more homogeneous valuation of the strategic options compared to the other firms.	*Accepted*
Hypothesis 6b: The firms in the main subsector of the territory offer a lower valuation of the need for public policies compared to the other firms.	*Partly*

subsector) seems to have an impact on the strategic analysis perceptions and that there is a certain influence of specialization on the perceptions of strategic needs and policy demands.

The results seem to indicate that the pressure of globalization upon traditional manufacturing leads firms to broaden their range of strategic responses beyond the more customary solutions. These proposals can be classified into myopic (short-run), defensive (restructuring) and offensive (multilocation, diversification towards high value-added subsectors or technological innovation) (Pla *et al.*, 2007).

The results also point towards interesting reflections with respect to the degree of generalization of the analysis and the level of specificity of the suggested actions. There is less variability in the home-textile subsector and the manufacturing specialization of the firms is an important determinant of their competitiveness level. Future studies carried out for the textile-clothing sector should take into account these aspects, as some of the main threats and opportunities diagnosed can be higher in some subsectors than in others. Along similar lines, some strategic options and policy actions that have been proposed as solutions to overcome the difficulties of the textile-clothing sector should consider, among other aspects, the specific competitive position of the firms in each subgroup (Dicken, 2003; Jones and Hayes, 2004; Buxey, 2005). This reorientation would improve their efficacy.

6.6 The subsector effect on international activity

Previous chapters have shown that, in the current situation of the textile-clothing sector, internationalization has been one of the key strategies for survival. Nevertheless, there is a paucity of studies looking at whether that influence can be extended to importing activities and whether there is some moderating role for the subsector. So, the objective pursued in this section is the analysis of the influence of the subsector on the internationalization strategy of the firms from the twin perspectives of exports and imports. Consequently, internationalization is seen as a dependent variable of the subsector to which the firm belongs. Hence, the international activity of a firm would be related to its manufacturing activity and the result would depend on whether the products manufactured were inputs or final goods, or whether that activity was directed towards mature or declining markets.

The internationalization variable was operationalized using the start of importing activity, the start of exporting activity, the propensity to import and the propensity to export. From the theory and the empirical results presented previously, we can test the proposed hypotheses on the possible influence that subsector specialization may have on the start and intensity of international activity:

> *Hypothesis 7*: The start of international activity will occur earlier for firms that are specialized in the main subsector in the district.

> *Hypothesis 8*: The intensity of international activity will differ for firms that are specialized in the main subsector in the district.

Given that in the analysis two types of international activity (imports and exports) have been distinguished, the two more general hypotheses should be disaggregated in four more specific hypotheses (7a, 7b, 8a and 8b) (Figure 6.9).

6.6.1 The subsector effect on the start of international activity

The statistical applicability obtained through the transformations carried out on international activity variables allows the parametric testing of hypotheses 7a and 7b. The results on the influence of the subsector are shown in Table 6.21. As can be seen, the start of importing occurs later, on average, for the main subsector in the territory (home textiles) (13.96 years) than in others (12.03 years). However, it seems that the tendency is inverted for the influence on exporting activity, as the start occurs sooner for the home subsector than for the others (10.75 against 11.29 years, respectively). This seems to be in line with the result that this activity is different according to the type of textile firm in question (Dicken, 2003; Buxey, 2005).

The issue now is the understanding of the dispersion of those scores on the transformed data among the various subgroups and of the statistical significance of the results. As shown by the four situations described (Figure 6.10), some values indicate the possibility of finding real differences in terms of centrality, symmetry and variability of the observations. However, it can be seen that none contains a set of values whose dispersion is very different. This finding increases the reliability of the results obtained (Hair *et al.*, 2000). Besides, the statistical results for the Levene test in Table 6.21 validate the respect for the conditions of applicability (Levene > 0.05 in all cases).

Finally, the F-test p-value (0.485 and 0.903) (Table 6.21) indicates that there are no real differences in the average scores achieved in testing the subsector effect. Hence, hypotheses 7a and 7b are not verified. This leads us to conclude that there does not seem to exist a significant influence of specialization on the start of international activity, whether on the import or on the export side.

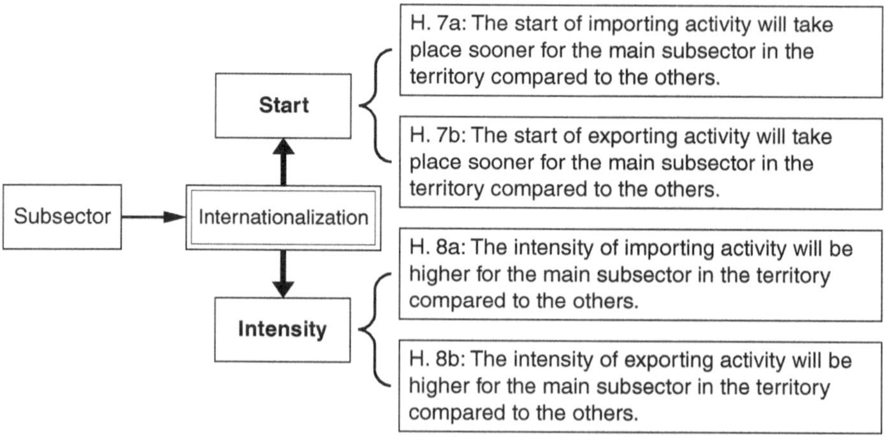

Figure 6.9 Diagrammatic representation of the subsector effect on internationalization.

Table 6.21 Summary of the subsector effect on the start of internationalization

	Averages		*Statistics*		*Number of cases (others/home)*
	Others	*Home*	*F-test*	*Levene*	
Years to start importing	12.03	13.96	0.485	0.925	40/48
Years to start exporting	11.29	10.75	0.903	0.588	51/65

6.6.2 The subsector effect on the intensity of international activity

Given that to analyse the intensity of international activity two types of international activities (imports and exports) have been distinguished, the more general hypothesis (the uneven influence on international intensity) is disaggregated into two more specific hypotheses (8a and 8b). The applicability obtained previously allows us to carry out a multivariate analysis. Table 6.22 shows that the result has the expected sign: the firms in the main subsector in the territory register, on average, a lower volume of purchases in international markets and, in turn, sell a higher percentage of their products in international markets.

The dispersion of transformed data is low, as shown in Figure 6.11 (high degree of reliability). With respect to F-test, hypothesis 8b seems to be met (p-value 0.007). The statistic associated with hypothesis 8a (p-value 0.137) does not allow its acceptance. Hence, we can conclude that the firms in the home subsector are, on average, exporting significantly more than those in the others group of subsectors, although we find no difference with respect to import intensity.

As a conclusion, Table 6.23 presents the results of the testing of hypotheses regarding the subsector effect on international activity. The main inference from this table is that the main activity of a firm (subsector) does not seem to have a high impact on imports, with mixed results for the exporting strategy.

In sum, the results of this section regarding the subsector effect on international activity seem to suggest that the intensity of importing activity does not

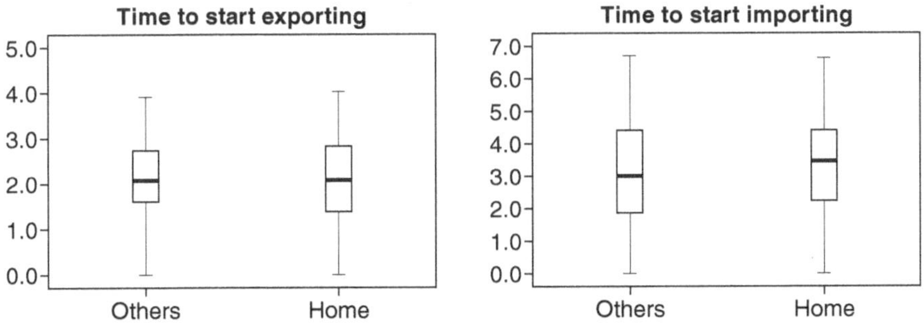

Figure 6.10 Box plot of the start of international activity according to the subsector.

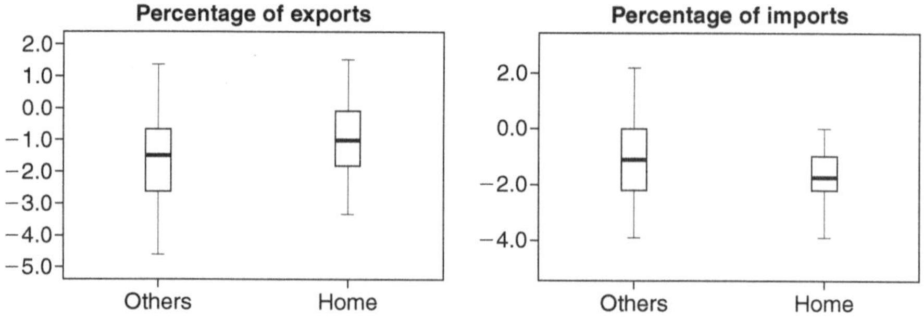

Figure 6.11 Box plot of the intensity of international activity according to the subsector.

seem to depend on the subsector. Although the sign of the results obtained seems to be in line with the expected influence (less imports in the main subsector), there is no empirical evidence to validate this hypothesis. On the contrary, the exporting intensity seems to depend on the subsector. According to the expected influence, the firms in the home subsector show percentages significantly above the rest of the firms.

The result of the more general analysis (start and intensity) reinforces somewhat the one obtained for the district effect on international activity. This means that, more than productive specialization, the main trigger for firms to become global is related to the territorial networks to which they belong. On the contrary, similar average percentages for the other subsectors have not supported an inhibiting effect against delocation, which would be expected (for example, for the knitting and clothing subsectors). Given the recent difficulties of the sector, this finding can point towards the process still being in its initial stages. These conclusions raise important questions relating to the conditions under which the district production model should evolve, namely with respect to the value chain.

Table 6.22 Summary of the subsector effect on the international intensity

	Averages		*Statistics*		*Number of cases (others/home)*
	Others	*Home*	*F-test*	*Levene*	
% imports on purchases	30.17	20.75	0.137	0.156	40/47
% exports on sales	30.53	39.20	*0.007	0.008	56/76

Notes

* significant at 5%. Although the exports–sales ratio variable was transformed to be analysed as a variable following a normal distribution, the results from the Levene test show that the variances are not homogeneous for the subgroups defined according to the subsector. As the non-parametric test also indicates significance at the 5% level (Mann-Whitney's U=0.015), we decided to continue the analysis with the transformed variable.

Table 6.23 Summary of hypothesis-testing on internationalization

Hypothesis 7: The start of international activity will take place sooner for the main subsector in the territory compared to the rest.	
Hypothesis 7a: The start of importing activity will take place sooner for the main subsector in the territory compared to the others.	*Rejected*
Hypothesis 7b: The start of exporting activity will take place sooner for the main subsector in the territory compared to the others.	*Rejected*
Hypothesis 8: The intensity of international activity will be different for the main subsector in the territory compared to the rest.	
Hypothesis 8a: The intensity of importing activity will be higher for the main subsector in the territory compared to the others.	*Rejected*
Hypothesis 8b: The intensity of exporting activity will be higher for the main subsector in the territory compared to the others.	*Accepted*

As a whole, the chapter has shown that the textile-clothing firms sampled have heterogeneous participation and do not fully exploit the advantages associated with geographical concentration. At the same time, the pressure of globalization increases the need for a higher degree of interconnectedness between firms and for utilization of shared resources within the district. The importance of a renewed strategy for the survival of those firms was made clear when distinguishing firms by their specialization and internationalization characteristics. This distinction is also important when implementing various kinds of policies.

7 Conclusions and remarks

The textile-clothing industry is the most representative of traditional manufacturing and it illustrates the challenges this labour-intensive sector is currently facing in Europe, compounded by reduced average firm size, weak technological development and geographical concentration. Traditional manufacturing is a mature sector in Europe, facing reduced demand and low innovation levels, together with growing competition from low-labour-cost countries, which in the case of textile-clothing has been fostered by the liberalization of its world trade since 2005 in a general context of globalization.

The textile-clothing industry is characterized by the frequent organization of its firms in geographical agglomerations in the search for external economies and production flexibility, of which the most well-known format is the territory known as the industrial district. The literature on industrial districts has highlighted their importance in the generation of external economies and competitive advantages for the firms locating there. The social and organizational characteristics of the district and its global orientation can help explain, to some extent, the managerial perceptions observed in the study carried out in the Comunidad Valenciana and analysed in Chapters 5 and 6. At the same time, the globalization process has been pointed out as a serious threat to industrial districts, to which they have responded with increased internationalization. The participation of district firms in international markets happens more through trade than through investment, both because most of them are SMEs and because the existence of the district is an incentive against the delocation of production.

Most strategic recommendations usually made to traditional manufacturing firms are of a general nature, typically of diversification towards non-traditional manufacturing. However, the analysis carried out in this book shows that this general view is not always correct. It is necessary to take into account the participation or not in an industrial district and the particular traditional manufacturing subsector in which each firm operates, or its subsector of specialization. The analysis presented has identified the Alcoi-Ontinyent industrial district as the most important textile-clothing concentration in Spain, and has compared the firms located in that district to firms outside it, finding some heterogeneity in the participation and exploitation of territorial advantages. At the same time, that district is very specialized in home textiles, and in the work presented it is not always possible to distinguish the results that are due to the district presence from those that are due to the specialization in home textiles. Nevertheless, the end result is that the strategic recommendations made to district firms that specialize in the district's main activity (home textiles) may not necessarily be the same as those for firms outside that district or with different specializations. Moreover, the differences found with respect to managerial perceptions and internationalization can be explained to a large extent by the district location and the specialization in the district's main activity.

7.1 Main conclusions

In general, the perception of structural crisis in the textile-clothing industry is not altered by the territory, indicating that the difficulties experienced by the sector in Europe are common to all SMEs. However, district firms are heterogeneous with respect to strategic needs and priorities attributed to public policies. This is because the population of firms forming an industrial district operates at different levels of the production process. Consequently, a particular corporate strategy, for example 'to explore new markets', is perceived as very necessary only by those firms operating in the final stages of the production process (downstream firms).

The district location also influences the start and intensity of exporting: district firms start exporting earlier and show higher export propensity. In the particular sample studied, this effect could occur through the subsector specialization, as the district's main specialization is home textiles, or it could be the case that the district's positive impact on exports is maintained independently of the specialization pattern. In the sample studied, home-textile firms do not start exporting earlier, although they show higher export intensity. It remains questionable whether the higher export intensity observed among home-textile firms in the district is due to their location or to their specialization. In order to disentangle these two effects, it is necessary to account for both simultaneously.

In the sample, firms in the main subsector of the territory (home textiles) do, however, differ from other textile-clothing firms as they have a more homogeneous perception of the environmental threats and strategic needs, a better competitive position, and show significant differences with respect to their prioritization of public policies. These differences may or may not be found in other districts where home textiles are not the main specialization.

The conclusions derived from testing the hypotheses proposed are summarized in Table 7.1. The results obtained improve current knowledge in a number of ways. First, they allow the improvement of strategic diagnoses and more efficacy in strategic recommendations. Second, they provide a view of the evolutionary dynamics of industrial districts. Third, they cast some light on the causes of heterogeneity on the participation and exploration of territorial advantages (subsector effect). Finally, they contribute towards a better understanding of two key aspects of the evolution of traditional manufacturing in Europe: the process of delocation and the diversification towards high value-added specializations.

Table 7.1 Summary of conclusions

Hypotheses	*Validation*
H. 1: District firms have more homogeneous and more positive managerial perceptions of the environment and competitive position than non-district firms.	Rejected
H. 2: District firms have more heterogeneous and more negative managerial perceptions of strategic needs and policy priorities.	Mixed results
H. 3: The start of international activity will occur earlier for district firms than for non-district firms.	Accepted
H. 4: The intensity of international activity will be different in the district and outside.	Accepted only for exports (higher export intensity for district firms)
H. 5: Firms operating in the main district subsector have more positive and more homogeneous perceptions of business performance.	Accepted with mixed results for environmental factors
H. 6: Firms operating in the main district subsector have more negative and more heterogeneous perceptions of strategic needs and policy priorities.	Accepted with mixed results for public policies
H. 7: The start of international activity will occur earlier for firms that are specialized in the main subsector in the district.	Rejected
H. 8: The intensity of international activity will differ for firms that are specialized in the main subsector in the district.	Accepted only for exports (higher export intensity for home-textile firms)

The closure and/or relocation of plants to low-labour-cost countries can be reduced by resorting to geographically concentrated networks, as these provide incentives to cluster and lower the attractiveness of distant locations. Moreover, recently proposed strategies such as innovation, flexibility, specialization or internationalization are actually facilitated by geographical concentration, especially in the case of SMEs, which receive tangible and intangible benefits from clustering.

7.2 Some caveats

The caveats that must be considered are mostly methodological and concern specifically the data used, the geographical delimitation of the industrial district and the classification of the firms into the textile-clothing subsectors.

The data are subject to some restrictions that depend on the measurement scales used, the geographical boundaries to which they are subject and the period of time to which they relate (Hair *et al.*, 2000). The use of five-point Likert scales in the questionnaire increases simplicity and objectivity in the treatment of the questionnaires, but there is always some subjectivity in the interpretation of the questions by the respondents, and some information is lost by forcing a quantification of perceptions. There may also be some bias introduced by sampling only firms affiliated with ATEVAL, although this bias may be reduced by considering that ATEVAL includes firms in the most important geographical concentration of the textile-clothing industry in Spain. Another source of bias is the fact that only 201 firms provided valid responses, although it has also been shown where this sample may differ or not from the population that originated it. Finally, the questionnaire was distributed in 2005, when the WTO liberalized world trade in textiles-clothing, and for that reason European firms, especially the SMEs, were particularly pessimistic about the evolution of the industry. It is conceivable that, four years later, some adjustment to the new realities has taken place, but as this book also highlights, this has probably followed different paths in industrial districts or outside, in large firms or in SMEs, or in the various textile-clothing subsectors that are more export-oriented or import-oriented.

The geographical delimitation of the industrial district is also subject to caveats, mainly derived from the industrial classification available (CNAE) and the use of CS. It is logical that for a municipality to be considered as part of an industrial district it should be more specialized in a certain industry than the average for the geographical unit of reference – that is, its CS should be higher than 1. However, the concept of high dependency on one particular economic activity led to the use of a higher threshold (CS > 10), which is to some extent arbitrary. But given that in the end the most important industrial district was identified as a case of analysis, the fixing of the threshold becomes irrelevant, and in practice simplifies the handling of data. Naturally, if the objective was to identify all the textile-clothing agglomerations in the country, including local production systems and clusters, a threshold of CS > 1 would be more adequate.

The classification of the firms into textile-clothing subsectors has been more subjective for those firms with activities in more than one subsector. Some firms did not declare one single subsector as its main activity and it was necessary to resort to ATEVAL, to experts and to our own judgement to classify them into only one of the subsectors.

7.3 The future

The extent to which the results obtained from this sample may be confirmed for other samples taken for firms with different characteristics, in different geographical locations and for different time periods, can only be confirmed by future studies. Nevertheless, this book has tried to provide a methodology of analysis that can be transposable to other samples and facilitate the comparison among future studies of European traditional manufacturing.

Future studies on the influence of the territory and specialization on globalized traditional manufacturing can improve the current proposals by addressing the caveats described previously and perhaps exploring to a greater extent the functional relationships established within the firms and the role of the managerial characteristics in determining perceptions and strategies. In future questionnaires, the characteristics of internationalization can be explored in more detail and complemented with information on firm productivity, profitability and growth, as well as the building of longitudinal series.

Finally, the book has shown that traditional manufacturing industries are heterogeneous and the subsectors within one particular industry need to be taken into account. As a consequence, besides the 'district effect' that is well-documented, the book proposes the 'subsector effect' as an additional moderating effect to be considered. From the joint consideration of both aspects, a new line of investigation can emerge that looks at the specialization–territory vector and its effect on firm performance.

The authors would like to encourage further studies of other traditional manufacturing sectors in other European countries in order to achieve a better understanding of how economic activities that are labour intensive and geographically concentrated have been responding to the double challenge of liberalization of world markets and continued European integration. Hopefully this book has provided an incentive and proposed a useful methodology of analysis.

Appendix 1

Questionnaire

IDENTIFICATION: NAME OF FIRM ______________________________

1.1. Gender: Male ☐ Female ☐

1.2. Age: __________

1.3. Status: Owner ☐ Manager ☐

1.4. Qualifications:
Primary school ______________________________
Secondary school ______________________________
Professional course ______________________________
University (1st cycle) ______________________________
University (beyond 1st cycle) ______________________________

1.5. Position in the firm ______________________________

1.6. How would you describe the current state of your firm?
good ☐ average ☐ poor ☐ bad ☐

1.7. Do you have affiliates abroad? yes ☐ no ☐

1.8. What year did the firm start exporting? ______________________

1.9 How are those exports evolving? growing ☐ static ☐ declining ☐

1.10 What is the share of sales currently due to exports? ____%

1.11. What year did the firm start importing? _______

1.12. How are those imports evolving? growing ☐ static ☐ declining ☐

1.13. What is the share of purchases currently due to imports? ________%

IMPORTANCE OF ENVIRONMENTAL STRATEGIC FACTORS

How is your firm being affected by the following environmental strategic factors? **Mark your reply using an 'x', with (1) affects very negatively, (2) negatively, (3) not important, (4) affects positively and (5) positively.**

TABLE A1

2.1	The dependency on foreign suppliers of machinery	1 2 3 4 5
2.2	The EU enlargement	1 2 3 4 5
2.3	Large retailers buying directly from emerging markets	1 2 3 4 5
2.4	Importers of finished products	1 2 3 4 5
2.5	The euro appreciating against the US dollar	1 2 3 4 5
2.6	The growing international competition from Asia (China, Pakistan, India, etc.)	1 2 3 4 5
2.7	The changes in consumption habits for textile products	1 2 3 4 5
2.8	The proximity to firms in all stages of the value chain	1 2 3 4 5
2.9	The continuous increase in oil prices	1 2 3 4 5
2.10	Most companies using very similar competitive strategies	1 2 3 4 5
2.11	The bargaining power of yarn manufacturers	1 2 3 4 5
2.12	Little political sensitivity to the problems faced by the textile industry	1 2 3 4 5
2.13	The absence of international harmonization of labels	1 2 3 4 5
2.14	The progressive disappearance of traditional retailers	1 2 3 4 5
2.15	The threat by the EU to impose safeguard clauses	1 2 3 4 5
2.16	The centralization of the purchase function carried out by many distributors	1 2 3 4 5

IMPORTANCE OF INTERNAL STRATEGIC FACTORS

Indicate the position of your firm when compared to its main competitors in Spain and the rest of Europe (Portugal, Germany, Belgium, Italy, etc) with respect to the following indicators. **Mark your reply using an 'x', with (1) much worse than its competitors, (2) somewhat worse, (3) similar, (4) somewhat better and (5) much better than its competitors.**

TABLE A2

3.1	Technical and productive flexibility	1 2 3 4 5
3.2	Result of innovative activity (design, packaging, reliability, etc.)	1 2 3 4 5
3.3	Firm size	1 2 3 4 5
3.4	Geographical location	1 2 3 4 5
3.5	Decentralized model for decision-making	1 2 3 4 5
3.6	Workers' flexibility	1 2 3 4 5
3.7	Workers' skill level	1 2 3 4 5
3.8	Reputation and brand	1 2 3 4 5
3.9	Product prices	1 2 3 4 5
3.10	Turnaround of samples and designs	1 2 3 4 5
3.11	Establishing and following a strategic plan	1 2 3 4 5
3.12	Capacity to adapt to customer requirements	1 2 3 4 5
3.13	Technical experience in the sector	1 2 3 4 5
3.14	Presence of products in international markets	1 2 3 4 5
3.15	Development achieved with strategies of cooperation	1 2 3 4 5

IMPORTANCE OF STRATEGIC NEEDS

Taking into account the current situation of your firm and the following strategies, indicate the current importance of each one for your firm, with **(1) not needed, (2) somewhat needed, (3) needed, (4) much needed and (5) very much needed.**

TABLE A3

4.1	Increase the size of the firm	1 2 3 4 5
4.2	Establish strategic alliances with competitors (cooperation)	1 2 3 4 5
4.3	Progressively abandon production activities and become a distributor-importer	1 2 3 4 5
4.4	Higher control of distribution (own shops or franchising)	1 2 3 4 5
4.5	Multilocalize or reorganize resources internationally (outward FDI)	1 2 3 4 5
4.6	Diversify towards new subsectors (technical textiles)	1 2 3 4 5
4.7	Explore new markets for their products (South Asia, Latin America or Eastern Europe)	1 2 3 4 5
4.8	Delocate the firm to another country	1 2 3 4 5
4.9	Specialize in a market niche	1 2 3 4 5
4.10	Increase the product range offered (textiles or otherwise)	1 2 3 4 5
4.11	Switch to another activity unrelated to textile	1 2 3 4 5
4.12	Establish agreements for international subcontracting	1 2 3 4 5
4.13	Reduce the number of customers to better specialize in one or a few of them	1 2 3 4 5

PRIORITY OF PUBLIC POLICIES

Taking into account the current situation of your firm and the following public policies, indicate the current priority of each one for your firm, with **(1) not needed, (2) needed in the long-run, (3) needed in the medium-run, (4) needed in the short-run and (5) needed urgently.**

TABLE A4

5.1	Facilitate entrepreneurial membership of institutions or associations	1 2 3 4 5
5.2	Increase economic support for lay-offs	1 2 3 4 5
5.3	International protection of brand and design	1 2 3 4 5
5.4	Increase the support to improve market access to international markets	1 2 3 4 5
5.5	Pursue international fairness in the rules of the game for competition	1 2 3 4 5
5.6	Improve the image of the national textile in international markets through a quality seal	1 2 3 4 5
5.7	Establish fiscal support to facilitate M&As	1 2 3 4 5
5.8	Higher financial support for diversification towards new textile products	1 2 3 4 5
5.9	Economic support for entrepreneurial cooperation with related sectors (furniture, lighting, etc.)	1 2 3 4 5
5.10	Impose heavier trade barriers to limit imports of textile products from outside the EU	1 2 3 4 5
5.11	Increase support for training of employees	1 2 3 4 5

6.1 What are your views on the future evolution of your firm?
improve ☐ static ☐ worsen ☐

Appendix 2

Statistical tables

Table A2.1 Criteria and number of factors in the rotation of the environment group

Sums of the square saturations of the extraction		
Total	*% of variance*	*% accumulated*
2.905	18.157	18.157
1.822	11.387	29.544
1.340	8.377	37.921
1.242	7.764	45.686
1.159	7.243	52.929

Table A2.2 Structural matrix of the environment group

	Component				
	1	*2*	*3*	*4*	*5*
The dependency on foreign suppliers of machinery	0.378	–0.283	–0.426	–0.361	0.226
The EU enlargement	0.124	–0.058	0.083	0.744	0.087
Large retailers buying directly from emerging markets	0.132	–0.652	0.036	0.379	0.220
Importers of finished products	0.243	–0.750	0.189	–0.051	0.179
The euro appreciating against the US dollar	–0.089	–0.166	0.671	0.114	0.168
The growing international competition from Asia (China, Pakistan, India, etc.)	–0.011	–0.769	0.101	–0.114	0.094
The changes in consumption habits for textile products	0.206	–0.215	0.566	–0.355	0.129
The proximity to firms in all stages of the value chain	0.536	–0.017	0.606	0.072	–0.017
The continuous increase in oil prices	0.454	0.138	–0.095	–0.332	0.314
Most companies using very similar competitive strategies	0.665	–0.110	0.042	–0.251	0.293
The bargaining power of yarn manufacturers	0.322	0.379	0.112	–0.302	0.492
Little political sensitivity to the problems faced by the textile industry	0.334	–0.344	0.127	–0.040	0.501
The absence of international harmonization of labels	0.157	–0.051	–0.088	0.226	0.710
The progressive disappearance of traditional retailers	0.006	–0.185	0.155	–0.091	0.663
The threat by the EU to impose safeguard clauses against imports from outside the EU	0.626	–0.013	–0.025	0.097	0.001
The centralization of the purchase function carried out by many distributors	0.607	–0.279	0.076	0.125	0.255

Table A2.3 Criteria and number of factors of the rotation in the strategy group

Sums of the square saturations of the extraction		
Total	*% of variance*	*% accumulated*
3.583	27.560	27.560
1.497	11.519	39.079
1.147	8.821	47.900
1.091	8.396	56.296

Table A2.4 Structural matrix for the strategy group

	Component			
	1	*2*	*3*	*4*
Increase the size of the firm	–0.029	0.034	–0.020	0.851
Establish strategic alliances with competitors (cooperation)	0.292	0.316	0.340	0.658
Progressively abandon production activities and become a distributor-importer	0.527	0.084	0.630	0.313
Higher control of distribution (own shops or franchising)	0.261	0.274	0.699	0.236
Multilocalize or reorganize resources internationally (outward FDI)	0.618	0.396	0.318	0.324
Diversify towards new subsectors (technical textiles)	0.282	0.687	0.213	0.188
Explore new markets for their products (South Asia, Latin America or Eastern Europe)	0.126	0.756	0.055	0.154
Delocate the firm to another country	0.763	0.269	0.168	0.100
Specialize in a market niche	–0.014	0.490	0.629	0.122
Increase the product range offered (textiles or otherwise)	–0.122	0.542	0.341	0.441
Switch to another activity unrelated to textile	0.774	–0.045	0.207	0.023
Establish agreements for international subcontracting	0.479	0.425	0.305	0.454
Reduce the number of customers to better specialize in one or a few of them	0.172	–0.076	0.671	0.039

Table A2.5 Descriptive statistics according to location

	Firms outside the district			*Firms inside the district*		
	Average	*N*	*Std dev.*	*Average*	*N*	*Std dev.*
The dependency on foreign suppliers of machinery	2.85	81	0.42	2.83	108	0.50
The EU enlargement	3.06	81	0.87	3.05	109	0.92
Large retailers buying directly from emerging markets	1.67	79	0.92	1.75	109	0.88
Importers of finished products	1.60	81	0.79	1.56	109	0.70
The euro appreciating against the US dollar	1.90	81	0.90	1.80	107	0.89
The growing international competition from Asia (China, Pakistan, India, etc.)	1.38	81	0.58	1.30	109	0.54
The changes in consumption habits for textile products	2.40	80	0.92	2.34	108	0.84
The proximity to firms in all stages of the value chain	3.05	74	0.92	3.02	104	0.90
The continuous increase in oil prices	1.84	81	0.70	1.92	108	0.74
Most companies using very similar competitive strategies	2.53	81	0.74	2.52	107	0.66
The bargaining power of yarn manufacturers	2.58	79	0.78	2.64	107	0.65
Little political sensitivity to the problems faced by the textile industry	1.44	80	0.69	1.44	109	0.62
The absence of international harmonization of labels	2.36	80	0.78	2.35	107	0.73
The progressive disappearance of traditional retailers	1.91	81	0.73	2.10	109	0.72
The threat by the EU to impose safeguard clauses against imports from outside the EU	3.24	80	1.14	3.15	103	1.12
The centralization of the purchase function carried out by many distributors	2.41	80	0.84	2.42	107	0.75
Technical and productive flexibility	3.50	88	0.96	3.38	110	0.74
Result of innovative activity (design, packaging, reliability, etc.)	3.23	88	0.75	3.27	108	0.73
Firm size	3.13	88	0.92	3.14	111	0.92
Geographical location	2.95	88	0.82	3.16	111	0.75

continued

Table A2.5 continued

	Firms outside the district			*Firms inside the district*		
	Average	*N*	*Std dev.*	*Average*	*N*	*Std dev.*
Decentralized model for decision-making	3.17	87	0.67	3.30	108	0.66
Workers' flexibility	3.09	87	1.04	3.03	112	0.91
Workers' skill level	3.08	87	0.77	3.12	112	0.78
Reputation and brand	3.56	88	0.95	3.19	111	0.81
Product prices	2.98	88	0.86	3.14	111	0.62
Turnaround of samples and designs	3.28	86	0.99	3.42	108	0.73
Establishing and following a strategic plan	3.01	87	0.84	3.06	110	0.69
Capacity to adapt to the customer's requirements	3.78	88	0.89	3.75	112	0.70
Technical experience in the sector	3.85	88	0.78	3.67	112	0.73
Presence of products in international markets	2.67	84	1.11	2.90	108	0.87
Development achieved with strategies of cooperation	2.56	82	0.93	2.74	106	0.83
Increase the size of the firm	1.94	87	1.14	1.75	111	1.10
Establish strategic alliances with competitors (cooperation)	2.34	85	1.22	2.47	111	1.20
Progressively abandon production activities and become a distributor-importer	2.64	88	1.30	2.32	111	1.24
Higher control of distribution (own shops or franchising)	2.70	87	1.46	2.39	109	1.35
Multilocalize or reorganize resources internationally (outward FDI)	2.25	87	1.14	2.05	107	1.15
Diversify towards new subsectors (technical textiles)	2.83	87	1.16	2.81	110	1.24
Explore new markets for their products (South Asia, Latin America or Eastern Europe)	3.11	87	1.32	3.21	108	1.23
Delocate the firm to another country	1.52	84	0.99	1.51	109	0.87

Specialize in a market niche	3.27	86	1.14	2.85	109	1.27
Increase the product range offered (textiles or otherwise)	3.19	86	0.99	2.99	109	1.27
Switch to another activity unrelated to textile	1.94	85	1.32	2.31	110	1.43
Establish agreements for international subcontracting	2.31	84	1.14	2.21	108	1.14
Reduce the number of customers to better specialize in one or a few of them	1.57	88	0.88	1.48	110	0.89
Facilitate entrepreneurial membership of institutions or associations	3.19	86	1.28	2.91	110	1.27
Increase economic support for lay-offs	3.45	85	1.47	3.47	112	1.45
International protection of brand and design	4.07	87	1.09	3.95	110	1.22
Increase the support to improve market access to international markets	4.06	88	1.11	4.07	109	1.05
Pursue international fairness in the rules of the game for competition	4.53	87	0.94	4.64	111	0.83
Improve the image of the national textile in international markets through a quality seal	4.10	88	1.06	3.92	110	1.14
Establish fiscal support to facilitate M&As	3.38	86	1.31	3.42	109	1.28
Higher financial support for diversification towards new textile products	3.86	87	1.09	4.02	110	1.03
Economic support for entrepreneurial cooperation with related sectors (furniture, lighting, etc.)	3.28	86	1.25	3.36	110	1.21
Impose heavier trade barriers to limit imports of textile products from outside the EU	4.18	88	1.26	4.38	112	1.13
Increase support for training of employees	3.93	86	1.07	3.59	112	1.13
Average score of the environment group	2.26	82	0.32	2.25	109	0.30
Average score of the internal group	3.19	88	0.44	3.21	112	0.40
Average score of the strategy group	2.43	88	0.57	2.34	112	0.64
Average score of the policy group	3.82	88	0.63	3.79	112	0.70

Table A2.6 Descriptive statistics according to subsector

	Other subsectors			*Home textiles*		
	Average	*N*	*Std dev.*	*Average*	*N*	*Std dev.*
The dependency on foreign suppliers of machinery	2.86	105	0.47	2.82	83	0.47
The EU enlargement	2.99	105	0.91	3.13	84	0.88
Large retailers buying directly from emerging markets	1.65	104	0.77	1.80	83	1.03
Importers of finished products	1.49	106	0.69	1.69	83	0.78
The euro appreciating against the US dollar	1.99	105	0.88	1.66	82	0.89
The growing international competition from Asia (China, Pakistan, India, etc.)	1.38	106	0.59	1.29	83	0.51
The changes in consumption habits for textile products	2.44	104	0.92	2.27	83	0.81
The proximity to firms in all stages of the value chain	2.96	100	0.86	3.12	77	0.96
The continuous increase in oil prices	1.86	106	0.75	1.91	82	0.69
Most companies using very similar competitive strategies	2.55	105	0.68	2.49	82	0.72
The bargaining power of yarn manufacturers	2.56	102	0.74	2.69	83	0.66
Little political sensitivity to the problems faced by the textile industry	1.38	105	0.64	1.51	83	0.65
The absence of international harmonization of labels	2.31	104	0.85	2.40	82	0.61
The progressive disappearance of traditional retailers	2.01	106	0.72	2.02	83	0.73
The threat by the EU to impose safeguard clauses against imports from outside the EU	3.13	101	1.23	3.25	81	0.98
The centralization of the purchase function carried out by many distributors	2.43	104	0.73	2.39	82	0.86
Technical and productive flexibility	3.41	107	0.85	3.47	89	0.85
Result of innovative activity (design, packaging, reliability, etc.)	3.18	106	0.73	3.35	88	0.74
Firm size	3.12	108	0.97	3.15	89	0.86
Geographical location	3.08	108	0.86	3.07	89	0.69
Decentralized model for decision-making	3.20	106	0.67	3.29	87	0.66
Workers' flexibility	3.06	109	0.98	3.05	88	0.96
Workers' skill level	3.02	108	0.75	3.21	89	0.79
Reputation and brand	3.39	108	0.91	3.31	89	0.89
Product prices	3.06	108	0.71	3.09	89	0.78
Turnaround of samples and designs	3.21	103	0.81	3.54	89	0.87
Establishing and following a strategic plan	3.03	107	0.71	3.07	88	0.83

Capacity to adapt to the customer's requirements	3.72	109	0.76	3.82	89	0.83
Technical experience in the sector	3.77	109	0.78	3.73	89	0.73
Presence of products in international markets	2.75	102	1.03	2.86	88	0.95
Development achieved with strategies of cooperation	2.55	102	0.90	2.79	84	0.85
Increase the size of the firm	1.91	106	1.17	1.73	90	1.05
Establish strategic alliances with competitors (cooperation)	2.47	104	1.28	2.36	90	1.11
Progressively abandon production activities and become a distributor-importer	2.33	107	1.37	2.61	90	1.15
Higher control of distribution (own shops or franchising)	2.36	105	1.39	2.74	89	1.40
Multilocalize or reorganize resources internationally (outward FDI)	2.14	103	1.19	2.15	89	1.12
Diversify towards new subsectors (technical textiles)	2.84	106	1.27	2.79	89	1.10
Explore new markets for their products (South Asia, Latin America or Eastern Europe)	3.06	103	1.30	3.31	90	1.22
Delocate the firm to another country	1.51	103	0.94	1.53	88	0.91
Specialize in a market niche	3.04	106	1.21	3.02	87	1.27
Increase the product range offered (textiles or otherwise)	3.09	104	1.22	3.08	89	1.09
Switch to another activity unrelated to textile	2.21	104	1.45	2.08	89	1.32
Establish agreements for international subcontracting	2.31	101	1.21	2.18	89	1.05
Reduce the number of customers to better specialize in one or a few of them	1.57	106	0.91	1.47	90	0.86
Facilitate entrepreneurial membership of institutions or associations	3.04	106	1.34	3.05	88	1.20
Increase economic support for lay-offs	3.38	106	1.42	3.54	90	1.49
International protection of brand and design	3.98	105	1.23	4.02	90	1.09
Increase the support to improve market access to international markets	4.02	106	1.14	4.11	89	1.00
Pursue international fairness in the rules of the game for competition	4.64	106	0.84	4.56	90	0.89
Improve the image of the national textile in international markets through a quality seal	3.91	107	1.20	4.16	89	0.93
Establish fiscal support to facilitate M&As	3.43	104	1.34	3.39	89	1.22
Higher financial support for diversification towards new textile products	3.98	106	1.12	3.93	89	0.97
Economic support for entrepreneurial cooperation with related sectors (furniture, lighting, etc.)	3.27	104	1.31	3.41	90	1.13
Imposing heavier trade barriers to limit imports of textile products from outside the EU	4.44	108	1.10	4.14	90	1.24
Increase support for training of employees	3.73	106	1.18	3.76	90	1.03
Average score of the environment group	2.24	106	0.30	2.27	84	0.32
Average score of the internal group	3.17	109	0.42	3.26	89	0.41
Average score of the strategy group	2.38	108	0.68	2.39	90	0.53
Average score of the policy group	3.80	108	0.69	3.82	90	0.65

Table A2.7 Labels of internal variables

TECFLEX	Technical and productive flexibility
INNOVA	Result of innovative activity (design, packaging, reliability, etc.)
SIZE	Firm size
LOCATE	Geographical location
DECENTR	Decentralized model for decision-making
WORKFLEX	Workers' flexibility
SKILL	Workers' skill level
REPUTACI	Reputation and brand
PRICE	Product prices
RENEW	Turnaround of samples and designs
STRPLAN	Establishing and following a strategic plan
REQUIREM	Capacity to adapt to the customer's requirements
TECHNEXP	Technical experience in the sector
PRESENIT	Presence of products in international markets
DEVCOOP	Development achieved with strategies of cooperation

Table A2.8 Labels of public policy variables

ASSOCIAT	Facilitate entrepreneurial membership of institutions or associations
SUPLAYOFF	Increase economic support for lay-offs
PROTECIT	International protection of brand and design
ACCESSIT	Increase the support to improve market access to international markets
FAIRGAME	Pursue international fairness in the rules of the game for competition
IMAGE	Improve the image of the national textile in international markets through a quality seal
SUPFIS	Establish fiscal support to facilitate M&As
SUPDIVER	Higher financial support for diversification towards new textile products
SUPCOOP	Economic support for entrepreneurial cooperation with related sectors (furniture, lighting, etc.)
TRADEBAR	Imposing heavier trade barriers to limit imports of textile products from outside the EU
SUPTRAIN	Increase support for training of employees

Notes

2 Networks and industrial districts

1 The concept of industrial district is frequently confused with those of technopolis, local production system, agglomeration, cluster, industrial area or *milieux*. Section 2.2 clarifies the concept of industrial district and only this concept will be used in the book.

2 A value chain is composed of several units (generally firms) related commercially, technically and financially, which define vertical flows (upstream and downstream) that create interdependence between those units (Johnson *et al.*, 2007).

3 The concept of the entrepreneur can refer to both those that create new firms and those that start new projects within existing firms (for example, entering new markets or introducing new processes).

4 The study of the geographical agglomeration of industry has a long tradition in what Stephen Meardon (2000) calls 'geographical economics': German location theory (von Thünen, 1826; Weber, 1929; Christaller, 1933; Lösch, 1939), regional science (Isard, 1956, 1960), economic geography (he points to the work of several geographers since the 1970s) and the 'new' economic geography (Krugman, 1991a). For the Spanish case, Fernández (1996) shows that agglomeration economies play a fundamental role in the location decisions of manufacturing firms.

5 For a comprehensive treatment of the modelization of agglomeration economies see Fujita *et al.* (2000).

6 See Marques (2008) for a comprehensive review of benefits and costs of agglomeration of industry in the European context.

7 See, for example, Altenburg and Meyer-Stamer (1999) on Latin America, Das (1998) on the floor-tile industrial district of Morbi in India, McCormick (1999) on Africa, Schmitz (1995) on shoe makers in Brazil, Rabelloti (1995) on Mexico, and Visser (1999) on Peru's clothing industry, among others.

8 The industrial district distinguishes itself from the general 'economic region' by having industry as its dominant activity (Becattini, 1992).

9 Law 5 of the Italian government 'on promotion of the SME', number 317, October 1991.

10 Although the data seem to imply lower productivity for district firms, it should be noted that industrial districts in Spain are very specialized in labour-intensive sectors, such as toys, textiles and furniture (Camisón, 1994; Giner and Santa María, 2002).

11 The lower degree of vertical integration is seen as specialization in particular stages of the production process. It can be measured, for each firm, by indicators such as the ratio between value added and total sales or operational revenues, or the share of inputs internally consumed.

12 Teece *et al.* (1994) suggested that an industry commonly comprises vertically integrated firms, specialized firms, diversified firms, hollow firms and networks. According

to this classification, Langlois and Robertson (1995) justify a lower level of vertical integration in a system characterized by specialized independent firms with a high degree of coordination of production due to strong cooperation relationships.

13 Tattara (2001) argues that the district is formed by many SMEs and microfirms with little transparency (double accounting and poor balance sheets) that maintain an informal system of relationships based on personal trust.

3 Globalization, internationalization strategy and industrial districts

1 The opening up of the EU eastwards was one of the major changes that took place in the economic environment since the fall of the Berlin Wall in 1989. Following 15 years of trade and investment liberalization, eight Central and Eastern European countries became EU members in May 2004 (Czech Republic, Estonia, Hungary, Latvia, Lithuania, Poland, Slovakia and Slovenia) and two more (Bulgaria and Romania) joined in January 2007. Currently Croatia and the Former Yugoslav Republic of Macedonia (FYROM) are candidates for EU membership.

2 The management literature distinguishes three strategic levels: corporate, business and functional. For the purposes of this chapter, only the corporate level will be discussed with respect to the firm's internationalization strategy. The strategies used at the business level and the resources invested in each functional area will not be analysed here.

3 Many firms start internationalizing by exporting, whether it is indirectly (passively) or directly (actively). In the first case, the exporting firm uses an intermediary located in the origin country that manages operations and is paid either a fixed amount or a variable fee. The intermediary can be a foreign buyer, agent or import–export representative, trading company distributor, or export consortium. In the case of direct export, the firm takes charge of all the administrative, logistic and financial aspects of the export operations, up to the final buyers in the export market. Also in this case the exporting firm may use its own commercial representatives, establish contracts with representatives in the export market that negotiate on the firm's behalf, or use local distributors or create its own commercial subsidiaries abroad. There is a consensus in the literature that the choice between one or another type of export, or the decision in favour of contractual models or FDI, depends on the degree of control over marketing strategies, the level of resources that can be committed to international activities and the potential level of benefits that can be obtained (Pla and León, 2004).

4 For two comprehensive surveys on why firms choose to export and invest see Helpman (2006) for the theoretical literature and Greenaway and Kneller (2007) for the empirics.

5 See Rauch (1999), Rauch and Trindade (2003), Zahra and George (2002) and Rialp and Rialp (2001) among others.

4 The textile-clothing industry in Europe: characterization and recent evolution

1 In May 2004, the Czech Republic, Estonia, Hungary, Latvia, Lithuania, Poland, Slovakia and Slovenia became members of the European Union. In January 2007 it was the turn of Bulgaria and Romania.

2 The OECD (1992) classifies manufacturing industry into three broad groups according to technological intensity: (1) high technological intensity (advanced sectors) – office machinery, electric machinery and instruments, precision instruments; (2) medium technological intensity (intermediate sectors) – agricultural and industrial machinery, transport equipment, chemicals, rubber and plastics; and (3) low technological intensity (traditional sectors) – food, beverages and tobacco, basic metals, other metallic products, minerals and non-metallic minerals, paper and printing, textile-clothing, leather and footwear.

3 The history of the textile-clothing industry has led to the formation of many industry associations. At the European level, one example is EURATEX (European Apparel and Textile Organization). In Spain, another example is ATEVAL (Association of Textile Entrepreneurs of the Comunidad Valenciana).

4 Eurostat compiles data on the consumption expenditure of EU citizens from national Household Expenditure Surveys. It considers 12 categories of expenditure, which here have been grouped into seven categories to facilitate the analysis: (1) food and beverages; (2) clothing and footwear; (3) housing and housing-related expenditure; (4) health and education; (5) transport and communications; (6) recreation and restaurantion; (7) other expenditure.

5 Eurostat applies NACE Rev. 2 to statistics provided from 2008. Given that the survey presented in the next two chapters was developed in 2005, NACE Rev. 1.1 is the most appropriate classification.

6 Including the construction sector.

7 The CS are calculated as: $CS_{ij} = \frac{E_{ij} / E_j}{E_i / E_n}$, where E_{ij} is employment in industry i in region j, E_j is total employment in region j, E_i is employment in industry i in the reference group (EU-27) and E_n is total employment in the reference group (EU-27). In our case the CS is the ratio of the share of textiles-clothing in a particular region's employment to its share in the EU-27.

8 The main trade partners outside Europe are the same for the EU-25 or EU-27. The analysis for the EU-15 has the advantage of highlighting the role of the Central and Eastern European member countries such as Romania and Bulgaria.

9 In Spain, microfirms are those employing 1–9 workers, small firms employ 10–49 workers, medium firms employ 50–249 workers and large firms are those with 250 workers or more.

10 We do not go beyond the NUTSIII (provincial) level, but see Boix and Galletto (2005) for a listing of Spanish textile-clothing concentrations at the municipal level.

11 This district is formed by eight municipalities (Agullent, Albaida, Alcoi, Banyeres de Mariola, Bocairent, Cocentaina, Muro de Alcoi and Ontinyent). It contains 33 per cent of the textile-clothing employment of the Comunidad Valenciana and has a specialization coefficient of 17.1 with respect to the Spanish average.

5 Methodology of analysis: an application to the Valencia textile-clothing industrial district

1 The LLM represent spatial units that are comparable geographically and statistically, and constitute spatially coherent systems of local interaction (Sforzi, 1992).

2 The light industrialization model is a system of organization of production based on the territorial concentration of a large number of SMEs specialized in one or a few stages of the production process.

3 For a summary see Becchetti *et al.* (2007).

4 The methodology of Lainé (2000) has also been applied by Santa María *et al.* (2004) to the footwear sector in Spain.

5 The CS used here is, as before, given by: $CS_{ij} = \frac{E_{ij} / E_j}{E_i / E_n}$. The numerator gives the share of an activity in the region's employment. The denominator gives the share of an activity in the country's employment.

6 The Herfindahl Index is given by: $HI_j = \sum_{i=1}^{n} \left(\frac{x_{ij}}{x_j} \right)^2$, where x_{ij} is the value of variable x (employment, investment, new firms) in sector i in territory j and x_j is the total value of that variable in territory j.

7 Its expression is similar to that of the CS, but refers to the territory rather than the sector.

8 Locational economies arise from the presence of firms in the same sector, while urbanization economies arise from the presence of firms in different sectors.

9 The two indexes are complementary: the CS measures the presence of a sector in a territory with respect to a reference territory, while the HI determines the degree of specialization of a territory with respect to the range of activities existing there.

10 Originally called *Censos de Población y Viviendas 2001*, it corresponds to the National Censuses carried out every ten years across Europe. It includes data on both population and housing. The 2001 Census considered all Spanish residents, whether of Spanish nationality or not, and all housing, whether private or institutional.

11 This information is provided by the AMADEUS database from Bureau Van Dijk.

12 Interestingly, the CS of the municipality of Arteixo (in the Galician province of A Coruña), where the Inditex group (which markets the brands Zara and Mango) is located, was only 7.9. The explanation is related to the fact that Inditex outsources a large part of its production, together with a high specialization of the municipality in other sectors of activity.

13 In Spanish, *Asociación de Empresarios Textiles de la Comunidad Valenciana.*

14 The sample error is the difference between the estimated and the true value of the proportion p; the confidence level is the probability of making a correct assessment regarding the posed hypotheses and is fixed at 95 per cent. The confidence interval obtained is given by: $\left[P-2\sqrt{\frac{p(1-p)}{n}}\sqrt{\frac{N-n}{N-1}},\ P+2\sqrt{\frac{p(1-p)}{n}}\sqrt{\frac{N-n}{N-1}}\right]$.

15 Microfirms are those with less than ten employees, small firms have 10–49 employees, medium firms have 50–249 employees and large firms are those with more than 250 employees.

16 For a summary on contingency tables and their applications, see Bryman and Bell (2007) and Bryman and Cramer (2004). Although in theory contingency tables could be applied to more than two variables, in practice their use is usually restricted to pairs of variables in order to facilitate the interpretation of the results.

17 A statistical package such as SPSS automatically returns corrected residuals in each cell of the contingency table. Each corrected residual returns the difference between the observed frequency of a given category and the frequency that would have been obtained in a coin flip (50/50 rule). See Bryman and Bell (2007) and Bryman and Cramer (2004) for a thorough explanation of the calculations and statistical package procedures.

18 A cell value is significant when it is above the 95 per cent critical value (±1.96) or at least above the 90 per cent critical value (±1.65).

19 See Eusebio and Llonch (2006) for a study of the importance of firm size for internationalization in the context of the Spanish textile-clothing industry.

6 Findings from the Valencia textile-clothing industrial district

1 It can be seen in the Bad/Worsening cell that the value of the corrected residuals (+3.0) indicates that the observation count is higher than would be expected for a 95 per cent confidence level.

2 The issue of safeguard clauses was brought up in the aftermath of the worldwide liberalization of trade in the textile-clothing sector as required by the WTO's Agreement on Textiles and Clothing, which had established 2005 as the end date for the system of quotas in place for many decades. Faced with a substantial rise in textile-clothing imports, mainly from China and other Asian countries, the EU considered invoking safeguard clauses, allowed by the WTO in cases of extreme and sudden disruption to the domestic market of the importing countries.

3 Previous analysis has shown this to be the geographical location of the largest textile-clothing district in Spain.
4 The reliability of these scales was measured using the Cronbach *alfa*. The scores obtained for each group were: environment (0.66), internal (0.78), strategies (0.77) and policies (0.8). According to Nunnally and Bernstein (1994), those results are acceptable.
5 This concept is used to describe the assumption that the variances of randomly drawn subsamples are not constant – that is, do not have the same dispersion. What is analysed here is whether some of the main sources of dispersion (managerial culture or competitive positioning) depend on location and main productive activity.
6 This technique requires that the following assumptions are met: (1) that the variables used meet the criteria of bivariate correlations higher than 0.3 (linearity); (2) the existence of an underlying structure among the variables analysed, for which we use the index of sample adequacy of Kaiser–Meyer–Oklin (KMO), the sphericity test of Bartlett and the determinant of the correlation matrix; and (3) the factor solutions obtained are simple and analysable (principles of parsimony and interpretability). The initial results from the statistical tests used indicate, for all groups, acceptable correlations among the variables and that, to a certain extent, the solutions can be interpreted sensibly.
7 According to Hair *et al.* (2000), the underlying assumptions of factor analysis are more conceptual than statistical. For that reason and due to the objective of the exercise (data reduction) the conditions of normality, homoscedasticity and linearity can be set aside.
8 If the assumption of normality had not been relaxed, other alternatives that could have been used instead of principal components analysis were: maximum likelihood estimation and non-linear principal components analysis.
9 The criterion used for calculating the factors to extract is based on auto values larger than 1, percentage of explained variance and the tail test. Given the nature of the data, the Oblimin rotation has been used to facilitate the interpretation of the factors or constructs. The tables laying out the sum of square saturations of the rotation and the *structure* matrices of both analyses are provided in Appendix 2 (Tables A2.1, A2.2, A2.3 and A2.4).
10 Although it is common to find research which relaxes the assumptions required for the use of the Ordinary Least Squares (OLS) regression method, here we have decided not to ignore that those assumptions are not met. First, because the assumptions were seriously not met for some items whose scores are situated in the boundaries. Second, because when codifying the ordinal scales from 1 to 5 a certain relationship was implicitly established that may not be exact. For example, the distance between 1 and 2 (say, comparing 'much worse than…' to 'somewhat worse than…') is not the same as between 2 and 3 (say, comparing 'somewhat worse than…' to 'average').
11 The descriptive statistics for each item (average, number of cases and standard deviation) are shown in Appendix 2 (Tables A2.5 and A2.6).
12 The Kolmogorov–Smirnov test is used to test the hypothesis that the independent variable follows a normal distribution.
13 The null hypothesis of the Levene test is that of equal variances across subsamples. The highest acceptable p-value threshold is 0.1 (confidence level of 90 per cent) so that when the test returns $p > 0.1$ the null hypothesis of equal variances can be accepted.
14 This model is explained in Greene (1999) and Agresti (2002). For a more advanced text, see Wooldridge (2002).
15 Given the ordinal answer variable (Y) with five points and the two independent variables (x_i) (location and subsector), we model the logit distribution function of Y conditional upon the values of x_i:

$$P(Y \leq j | x) = \log \frac{P(Y \leq j \mid x)}{P(Y > j \mid x)} = \alpha_j + \sum_{i=1}^{2} \beta_i x_i,\ j = 1,\ ...,\ 5,\ -1.$$

16 The various labels placed on each variable to simplify its reading are listed in Appendix 2 (Tables A2.7 and A2.8).

17 If the coefficient β is negative, the cumulative logit and the cumulative probabilities grow when the associated variable *x* grows. This means that the lower values of *Y* carry a higher probability, or alternatively that higher values carry a lower probability. If the coefficient is positive, lower (higher) values carry a lower (higher) probability (Agresti, 2002).

18 The box length is the distance between the first and third quartiles (50 per cent of the observations). The black line inside each box shows the position of the median, and the lines coming out of each box represent the distance between the maximum and the minimum (Hair *et al.*, 2000).

Bibliography

Abernathy, F., Volpe, A. and Weil, D. (2006) 'The future of the apparel and textile industries: prospects and choices for public and private actors', *Environment and Planning A*, 38 (12), 2207–32.

Agresti, A. (2002) *Categorical Data Analysis*, New York: Wiley.

Aitken, B., Hanson, G. and Harrison, A. (1997) 'Spillovers, foreign investment, and export behavior,' *Journal of International Economics*, 43 (1–2), 103–32.

Akerlof, G. and Kranton, R. (2000) 'Economics and identity,' *Quarterly Journal of Economics*, 115 (3), 715–53.

Aldrich, H. and Zimmer, C. (1985) 'Entrepreneurship through social networks', in D. L. Sexton and R. W. Smilor (eds), *The Art and Science of Entrepreneurship*, Cambridge, MA: Ballinguer Publishing Co.

Altenburg, T. and Meyer-Stamer, J. (1999) 'How to promote clusters: policy experiences from Latin America', *World Development*, 27 (9), 1693–713.

AMADEUS (2008) Bureau Van Dijk's database on European firms.

Amighini, A. and Rabellotti, R. (2006) 'How do Italian footwear industrial districts face globalisation?', *European Planning Studies*, 14 (4), 485–502.

Amin, A. (1993) 'The globalization of economy: an erosion of regional networks', in G. Grabher (ed.), *The Embedded Firm: On Socioeconomics of Industrial Networks*, London: Routledge.

Amin, A. and Robins, K. (1992) 'Distritos industriales and desarrollo regional: límites and posibilidades', in F. Pyke, G. Becattini and W. Sengenberger (ed.) *Los distritos industriales and las pequeñas empresas*, Madrid: Ministerio de Trabajo y Seguridad Social.

Annen, K. (2001) 'Inclusive and exclusive social capital in the small-firm sector in developing countries', *Journal of Institutional and Theoretical Economics*, 157 (2), 319.

Aspelund, A. and Moen, O. (2005) 'Small international firms: typology, performance and implications', *Management International Review*, 45, 37–57.

Bagella, M. (2000) 'The district advantage in small–medium firm internationalisation', in M. Bagella and L. Becchetti (eds), *The Competitive Advantage of Industrial Districts*, New York: Physica-Verlag.

Bagella, M., Becchetti, L. and Sacchi, S. (1998) 'Aglomerazione geografica delle imprese e performance nell'export: un'analisi empirica su microdati per l'Italia', *Sviluppo Locale*, 5 (8), 122–48.

Baldwin, R. and Forslid, R. (1997) 'The core–periphery model and endogenous growth', CEPR Discussion Paper 1749, London: CEPR.

Baldwin, R., Forslid, R., Haaland, J. and Midelfart-Knarvik, K. (2000) 'EU integration and outsiders: a simulation study of industrial location', Norwegian School of Economics and Business Administration Discussion Paper 02/2000.

Baldwin, R., Forslid, R., Martin, P., Ottaviano, G. and Robert-Nicoud, F. (2003) *Economic Geography and Public Policy*, Princeton, NJ: Princeton University Press.

Baptista, R. and Swann, P. (1998) 'Do firms clusters innovate more?', *Research Policy*, 27, 525–40.

Basevi, G. and Ottaviano, G. (2001) 'The district and the global economy: exporting versus foreign location', CEPR Discussion Paper 2976, London: CEPR.

Beamish, P. W. (1990) 'The internationalisation process for smaller Ontario firms: a research agenda', in A. M. Rugman (ed.), *Research for the Twenty-First Century: Canada's New Research Agenda*, Greenwich: JAI Press Inc.

Becattini, G. (1979) 'Dal "settore" industriale al "distretto" industriale: Alcune considerazioni sull'unitá di indagine dell'economia industriale', *Rivista di Economia e Politica Industriale*, 1, 7–21.

Becattini, G. (1992) 'El distrito industrial marshalliano como concepto socio-económico', in F. Pyke, G. Becattini and W. Sengenberger (eds), *Los distritos industriales and las pequeñas empresas*, Madrid: Ministerio de Trabajo y Seguridad Social.

Becattini, G. (2002) 'Industrial sectors and industrial districts: tools for industrial analysis', *European Planning Studies*, 10 (4), 483–93.

Becattini, G. (2004) *Industrial Districts: A New Approach to Industrial Change*, Cheltenham: Edward Elgar.

Becchetti, L. and Rossi, S. (2000a) 'EU and non EU export performance of Italian firms: is there an industrial district effect?', in M. Bagella and L. Becchetti (eds), *The Competitive Advantage of Industrial Districts*, New York: Physica-Verlag.

Becchetti, L. and Rossi, S. (2000b) 'The positive effect of industrial district on the export performance of Italian firms', *Review of Industrial Organization*, 16, 53–68.

Becchetti, L., de Panizza, A. and Oropallo, F. (2007) 'Role of industrial district externalities in export and value-added performance: evidence from the population of Italian firms', *Regional Studies*, 41 (5), 601–21.

Bell, G. (2005) 'Clusters, networks and firm innovativeness', *Strategic Management Journal*, 26, 287–95.

Bellandi, M. (2002) 'Italian industrial district: an industrial economics interpretation', *European Planning Studies*, 10 (4), 425–37.

Belloi, M. (2006) 'Distretti industriali: del locale al globale'. Online, available: http://biblioeco.unimo.it/ext/bibliografie/bibliografia_stampa.rtf (accessed 31 August 2009).

Belussi, F. and Caldari, K. (2009) 'At the origin of the industrial district: Alfred Marshall and the Cambridge school', *Cambridge Journal of Economics*, 3 (2), 335–55.

Belussi, F. and Sedita, S. R. (2009) 'Life cycle vs. multiple path dependency in industrial districts', *European Planning Studies*, 17 (4), 505–28.

Bernard, A. and Jensen, B. (2004) 'Why some firms export', *Review of Economics and Statistics*, 86 (2), 561–9.

Bilkey, W. J. and Tesar, G. (1977) 'the export behavior of smaller sized Wisconsin manufacturing firms', *Journal of International Business Studies*, 8, 93–8.

Boix, R. and Galletto, V. (2005) 'Sistemas locales de trabajo and distritos industriales Marshallianos en España', *Documents de Treball*, Departament d'Economia Aplicada, Universitat Autònoma de Barcelona.

Bramanti, A. and Senn, L. (1993) 'Entrepreneurs, firm, "milieu": three different specifications of networking activities', in D. Maillant, M. Quevit and L. Senn (eds), *Reseaux d'innovation and Milieux Innovateurs: Un Pari pour le Developpement Régional*, Neuchâtel: GREMI-EDES.

Brenner, T. (2003) 'An identification of local industrial clusters in Germany', *Papers on Economics & Evolution*, Max Planck Institute. Online, available: www.mpiew-jena.mpg.de/english/staff/WP0304.pdf (accessed 31 August 2009).

Brenner, T. (2004) 'Industrial districts: a tipology from an evolutionary perspective', paper presented at the DRUID Summer Conference.

Brulhart, M., Crozet, M. and Koenig, P. (2004) 'Enlargement and the EU periphery: the impact of changing market potential', *World Economy*, 27, 853–76.

Brusco, S. (1992) 'El concepto de distrito industrial: su génesis', in F. Pyke, G. Becattini and W. Sengenberger (eds), *Los distritos industriales and las pequeñas empresas*, Madrid: Ministerio de Trabajo y Seguridad Social.

Bryman, A. and Bell, E. (2007) *Business Research Methods*, Oxford: Oxford University Press.

Bryman, A. and Cramer, D. (2004) *Quantitative Data Analysis with SPSS: A Guide for Social Scientists*, London: Routledge.

Buckley, P. J. and Ghauri, P. N. (2004) 'Globalisation, economic geography and the strategy of multinational enterprises', *Journal of International Business Studies*, 35 (2), 81–98.

Buxey, G. (2005) 'Globalisation and manufacturing strategy in the TCF industry', *International Journal of Operations & Production Management*, 25 (2), 100–13.

Cainelli, L. and De Liso, N. (2004) 'Can a Marshallian industrial district be innovative? The case of Italy', in G. Cainelli and R. Zoboni (eds), *The Evolution of Industrial Districts*, New York: Physica-Verlag.

Cairncross, F. (1997) *The Death of Distance: How the Communications Revolution Will Change our Lives*, Boston, MA: Harvard Business School Press.

Camisón, C. (1994) 'Dirección de empresas en entornos globales and abiertos: hacia la estrategia de alianzas', *Revista Europea de Dirección y Economía de la Empresa*, 2 (3), 9–37.

Camisón, C. (2001) *La competitividad de la empresa industrial de la Comunidad Valenciana: análisis del efecto del atractivo del entorno, los distritos industriales and las estrategias empresariales*, Valencia: Tirant Lo Blanch.

Camisón, C. (2004) 'Shared, competitive, and comparative advantages: a competence-based view of industrial district competitiveness', *Environment and Planning A*, 36 (12), 2227–56.

Camisón, C. and Guía, J. (2001) 'Integración vertical, cooperación interempresarial and resultados: el efecto contingente de la coordinación implícita en un distrito industrial', *Cuadernos de Economía and Dirección de la Empresa*, 8, 51–76.

Capecchi, V. (1992) 'Un caso de especialización flexible: los distritos industriales de la Emilia-Romagna', in F. Pyke, G. Becattini and W. Sengenberger (eds), *Los distritos industriales and las pequeñas empresas*, Madrid: Ministerio de Trabajo y Seguridad Social.

Casavola, P., Pellegrini, G. and Romagnano, E. (2000) 'Imprese e mercato del lavoro nei distretti industriali', in L. F. Signorini (ed.), *Lo sviluppo locale: Un'indagine della Banca d'Italia sui distretti industriali*, Roma: Meridiana Libri.

Casson, M. (1998) 'Entrepreneurial networks: a theoretical perspective', in M. Moss and A. Slaven (eds), *Entrepreneurial Networks and Business Culture*, Sevilla: Fundación Fomento de la Historia Económica.

Cavusgil, S. T. (1980) 'On the internationalization process of firms', *European Research*, 8 (9), 273–81.

Cerejeira, J. C. (2002) 'Identification of the Portuguese industrial districts', Working Paper 17, Núcleo de Investigação em Microeconomia Aplicada (NIMA), Universidade do Minho.

CEUC (Conselleria d'Empresas, Universitat i Ciencia) (2006) *Plan de competitividad del sector textil*. Online, available: www.impiva.es/planescompetitividadsectorial/recursos/plan%20de%20competitividad%20textil.pdf (accessed 31 August 2009).

Chang, S. J. and Sing, H. (2000) 'Corporate and industry effects on business unit competitive position', *Strategic Management Journal*, 21, 739–52.

Chen, M. J., Su, K.-H. and Tsai, W. (2007) 'Competitive tension: the awareness–motivation–capability perspective', *Academy of Management Journal*, 50 (1), 101–18.

Chetty, S. and Blankenbourg-Holm, D. (2000a) 'The role of business networks in the internationalisation of manufacturing firms: a longitudinal case study', in R. B. MacNaughton (ed.), *Developments in Australian Marketing*, Stamford, CT: JAI Press Inc.

Chetty, S. and Blankenbourg-Holm, D. (2000b) 'Internationalisation of small to medium-size manufacturing firms: a network approach', *International Business Review*, 9, 77–93.

Chevassus-Lozza, E. and Galliano, D. (2003) 'Local spillovers, firm organization and export behavior: evidence from the French food industry', *Regional Studies*, 37 (2), 147–58.

Christaller, W. (1933) *Central Places in Southern Germany*, London: Prentice Hall.

Christoffersen, S. (2002) 'The textile industry: does R&D deliver success?', *Critical Review*, 12 (2), 77–95.

Clerides, S., Lach, S. and Tybout, J. (1998), 'Is learning by exporting important? Micro-dynamic evidence from Colombia, Mexico, and Morocco', *Quarterly Journal of Economics*, 113 (3), 903–47.

Comext (2009) Eurostat's trade database.

Comtrade (2009) United Nations' trade database.

Cooke, P. and Wills, D. (1999) 'Small firms, social capital and the enhancement of business performance through innovation programmes', *Small Business Economics*, 13 (3), 219–34.

Cooke, P., Clifton, N. and Oleaga, M. (2005) 'Social capital, firm embeddedness and regional development', *Regional Studies*, 39 (8), 1065–77.

Costa, M. T. (1988) 'Descentramiento productivo y difusión industrial: el modelo de especialización flexible', *Papeles de Economía Española*, 35, 251–76.

Costa, M. T. and Viladecans, E. (1999) 'The district effect and the competitiveness of manufacturing companies in local productive system', *Urban Studies*, 36 (12), 2085–98.

Courlet, C. and Pecqueur, B. (1994) 'Les systèmes industriels localisés en France: un nouveau modèle de développement', in G. Benko and A. Lipietz (eds), *Les régions qui gagnent: districts et réseaux: les noveaux paradigmes de la géographie économique*, Paris: Presses Universitaries de France.

Coviello, N. E. and McAuley, A. (1999) 'Internationalisation and the smaller firm: a review of contemporary empirical research', *Management International Review*, 39 (3), 223–56.

Crestanello, P. and Dalla Libera, P. E. (2003) 'International delocalisation of production: the case of the fashion industry of Vicenza', Paper presented at the Conference in Honour of Professor S. Brusco. Online, available: www.economia.unimo.it/convegni_seminari/CG_sept03/naviga.html (accessed 31 August 2009).

Cristovam, M. L. (2006). 'Restructuring case study: the textile and clothing industry in Portugal', Seminar on the role of public authorities in anticipating and managing labour market effects of economic change and restructuring, Directorate-General for Employment, Social Affairs and Equal Opportunities, European Commission.

Curran, J., Jarvis, R., Blackburn, R. A. and Black, S. (1993) 'Networks and small firms: constructs, methodological strategies and some findings', *International Small Business Journal*, 11 (2), 13–25.

Dalziel, M. (2007) 'A systems-based approach to industry classification', *Research Policy*, 36, 1559–74.

Dana, L. P., Hamilton, R. T. and Pauwells, B. (2007) 'Evaluating offshore and domestic production in the apparel industry: the small firm's perspective', *Journal of International Entrepreneurship*, 3–4 (5), 47–63.

Dangayach, G. S. and Deshmukh, S. G. (2001) 'Manufacturing strategy: literature review and some issues', *International Journal of Operations & Production Management*, 21 (7), 884–932.

Daniels, J., Radebaugh, L. and Sullivan, D. (2007) *Internatinal Business: Environments and Operations*, 11th edition, Upper Saddle River, NJ: Prentice Hall.

Das, K. (1998) 'Collective efficiency and firm strategy: study of an Indian industrial cluster', *Entrepreneurship & Regional Development*, 10 (1), 33–49.

Deardorff, A. and Stern, N. (2002) 'What you should know about globalization and the World Trade Organization', *Review of International Economics*, 10 (3), 404–23.

Dei Ottati, G. (1988) 'El mercado comunitario', *Papers de Seminari*, 29–30, 180–218.

Dei Ottati, G. (1994) 'The industrial district as a system and the balance between cooperation and competition', *European Planning Studies*, 2 (4), 463–83.

Dei Ottati, G. (2002) 'Social concertation and local development: the case of industrial district', *European Planning Studies*, 10 (4), 449–66.

DeMartino, R., McHardy Reid, D. and Zygliodopoulos, S. C. (2006) 'Balancing localization and globalization: exploring the impact of firm internationalization on a regional cluster', *Entrepreneurship & Regional Development*, 18, 1–24.

Denis, C., McMorrow, K. and Roger, W. (2006) *Globalisation: Trends, Issues and Macro Implications for the EU*, Brussels: European Commission, Directorate-General for Economic and Financial Affairs.

Detrell, J. (1998) 'El sector de tèxtils d'ús tècnic', *Debats Tecnològics*, 7, 28–42.

Devaraj, S., Hollingwort, D. G. and Schroeder, R. G. (2004) 'Generic manufacturing strategies and plant performance', *Journal of Operations Management*, 22, 313–33.

Devinney, T. M., Richard, P. J., Yip, G. S. and Johnson, G. (2005) *Measuring Organizational Performance in Management Research: A Synthesis of Measurement Challenges and Approaches*. Online, available: http://ssrn.com/abstract=814285 (accessed 31 August 2009).

Dicken, P. (2003) *Global Shift: Reshaping the Global Economic Map in the 21st Century*, London: SAGE Publications.

Dunford, M. (2006) 'Industrial districts, magic circles and the restructuring of the Italian textiles and clothing chain', *Economic Geography*, 82, 27–59.

Dunning, J. H. (1981) *International Production and the Multinational Enterprise*, London: Allen & Unwin.

Dunning, J. H. (1988) *Explaining International Production*, London: Unwin Hyman.

Dunning, J. H. (1992) *Multinational Enterprises and the Global Economy*, New York: Addison-Wesley.

Ebers, M. and Jarillo, J. C. (1998) 'The construction, forms, and consequences of industry networks', *International Studies of Management & Organization*, 27 (4), 3–21.

Elo, M. (2004) 'SME internationalisation process from a network perspective: a theoretical review', paper presented at the 30th EIBA Conference, Ljubljana, 5–8 December.

European Commission (2009) *Textiles Homepage*. Online, available: http://ec.europa.eu/enterprise/textile/index_en.htm (accessed 14 March 2009).

Eurostat (2009) Various statistical databases from Eurostat. Online, available: http://eurostat.ec.europa.eu (accessed 31 August 2009).

Eusebio, R. and Llonch, J. (2006) 'Los determinantes de la intensidad exportadora: un análisis comparativo entre empresas españolas e italianas del sector textil-confección', *Cuadernos de Economía y Dirección de Empresas*, 26, 95–122.

Fabiani, S., Pellegrini, G., Romagnano, E. and Signorini, L. F. (1998) 'L'efficienza delle imprese nei distretti industriali italiani', *Svilupo Locale*, 5 (9), 42–73.

Feenstra, R. (1998) 'Integration of trade and disintegration of production in the global economy', *Journal of Economic Perspectives*, 12 (4), 31–50.

Feenstra, R. and Hanson, G. (1996) 'Globalization, outsourcing, and wage inequality', *American Economic Review*, 86 (2), 240–5.

Fernández, V. (1996) 'Determinantes de la localización de las empresas industriales en España: nuevos resultados', *Economía Industrial*, 308, 149–61.

Fernández, Z. (1993) 'La organización interna como ventaja competitiva para la empresa', *Papeles de Economía Española*, 56, 178–93.

Fernández, Z. (1994) 'La banca universal: un futuro en entredicho', *Papeles de Economía Española*, 58, 262–75.

Fiegenbaum, A. and Karnani, A. (1991) 'Output flexibility: a competitive advantage for small firms', *Strategic Management Journal*, 12, 101–14.

Filatochev, I. and Toms, S. (2003) 'Corporate governance, strategy and survival in a declining industry: a study of UK cotton textile companies', *Journal of Management Studies*, 40, 895–920.

Fischer, S. (2003) 'Globalization and its challenges', *American Economic Review*, 93 (2), 1–30.

Fletcher, R. (2001) 'A holistic approach to internationalisation', *International Business Review*, 10 (1), 25–49.

Fujita, M., Krugman, P. and Venables, A. (2000) *The Spatial Economy: Cities, Regions and International Trade*, Cambridge, MA: MIT Press.

García, E. (1996) 'El estudio de las alianzas y relaciones interorganizativas en la dirección de empresas: tendencias actuales', *Revista Europea de Dirección y Economía de la Empresa*, 5 (3), 109–32.

Ghoshal, S. and Bartlett, C. A. (1990) 'The multinational corporation as an interorganizational network', *Academy of Management Review*, 15 (4), 603–25.

Giner, J. M. and Santa María, M. J. (2002) 'Territorial systems of small firms in Spain: an analysis of productive and organizational characteristics in industrial districts', *Entrepreneurship & Regional Development*, 14, 211–28.

Giuliani, E. (2007) 'Networks and heterogeneous performance or cluster firms', in K. Frenken (ed.), *Applied Evolutionary Economics and Economic Geography*, Cheltenham: Edward Elgar.

Goglio, S. (2002) 'Introduction: the industrial district as a proving ground', *European Planning Studies*, 10 (4), 421–4.

Golf, E., Molina, M. J., Pla, J. and Puig, F. (2005) *Plan Estratégico Textil de las Comarcas Centrales Valencianas*, Ontinyent: Proselección.

Gordon, I. and McCann, P. (2000) 'Industrial clusters: complexes, agglomeration and/or social networks?', *Urban Studies*, 37 (3), 513–32.

Grandinetti, R. and Passon, M. (2004) 'La situazione e le prospettive del distretto della sedia', *Camera di Commercio Industria Artigianato e Agricultura*. Online, available: www.promosedia.it/pdf/distretto_sedia.pdf (accessed 31 August 2009).

Grandinetti, R., Nassimbeni, G. and Sartor, M. (2004) 'The manufacturing internationalization of the Italian chair district', Proccedings of the 20th EGOS Conference.

Grandori, A. and Soda, G. (1995) 'Inter-firm networks: antecedents, mechanisms and forms', *Organization Studies*, 16 (2), 183–214.

Granovetter, M. S. (1985) 'Economic action and social structure: the problem of embeddedness', *American Journal of Sociology*, 91, 481–510.

Greenaway, D. and Kneller, R. (2007), 'Firm heterogeneity, exporting and foreign direct investment', *Economic Journal*, 117 (517), 134–61.

Greene, W. H. (1999) *Econometric Analysis*, Upper Saddle River, NJ: Prentice Hall.

Guercini, S. (2004) 'International competitive change and strategic behaviour of Italian textile-apparel firms', *Journal of Fashion Marketing and Management*, 8 (3), 320–39.

Gulati, R. and Gargiulo, M. (1999) 'Where do interorganizational networks come from?', *American Journal of Sociology*, 104 (5), 1439–93.

Gulati, R., Nohria, N. and Zaheer, A. (2000) 'Strategic networks', *Strategic Management Journal*, 21, 203–15.

Hadjikhani, A. and Ghauri, P. N. (2001) 'The behaviour of international firms in socio-political environments in the European Union', *Journal of Business Research*, 53 (3), 263–75.

Hadley, R. H. and Wilson, H. (2003) 'The network model of internationalisation and experiential knowledge', *International Business Review*, 12, 697–717.

Hair, J. F., Anderson, R. E., Tatham, R. L. and Black, W. C. (2000) *Multivariate Analysis*, London: Prentice Hall.

Hambrick, D. C. and Mason, P. A. (1984) 'Upper echelons: the organization as a reflection of its top managers', *Academy of Management Review*, 9 (2), 193–206.

Harris, R. G. (1993) 'Globalization, trade, and income', *Canadian Journal of Economics*, 26 (4), 755–76.

Harrison, B. (1994) 'The Italian industrial district and the crisis of the cooperative form: part I', *European Planning Studies*, 2 (1), 3–17.

Hayes, R. H. and Wheelwright, S. C. (1984) *Restoring Our Competitive Edge: Competing Through Manufacturing*, New York: Wiley.

Hawawini, G., Subramanian, V. and Verdin, P. (2003) 'Is performance driven by industry or firm specific factors? A new look to the evidence', *Strategic Management Journal*, 19, 211–19.

Head, K. and Mayer, T. (2006) 'Regional wage and employment responses to market potential in the EU', *Regional Science and Urban Economics*, 36, 573–94.

Helpman, E. (2006) 'Trade, FDI and the organization of firms', *Journal of Economic Literature*, 44, 589–630.

Helpman, E. and Krugman, P. (1985) *Market Structure and Foreign Trade*, Cambridge, MA: MIT Press.

Hernández, J. M., Fontrodona, J. and Pezzi, A. (2005) 'Mapa dels sistemes productius locals industrials a Catalunya', *Papers d'Economia Industrial*, 21, Generalitat de Catalunya.

Hite, J. M. and Hesterly, W. S. (2001) 'The evolution of firm networks: from emergence to early growth of the firm', *Strategic Management Journal*, 22, 275–86.

Hitt, M. A., Ireland, R. D. and Hoskisson, R. E. (2007) *Strategic Management: Competitiveness and Globalization: Concept and Cases*, n.p.: South-Western College Publishing.

Hofstede, G. (2001) *Culture's Consequences, Comparing Values, Behaviors, Institutions, and Organizations across Nations*, Thousand Oaks, CA: Sage Publications.

Human, S. E. and Provan, K. G. (1997) 'An emergent theory of structure and outcomes in small-firm strategic manufacturing networks', *Academy of Management Journal*, 40 (2), 368–403.

INE (Instituto Nacional de Estadística) (2004) *Censos de Población y Vivienda 2001*. Online, available: www.ine.es/censo/es/inicio.jsp (accessed 31 August 2009).

INE (Instituto Nacional de Estadística) (2005) *Directorio Central de Empresas: Demografía Industrial*. Online, available: www.ine.es (accessed 31 August 2009).

Inkpen, A. C. and Tsang, E. W. (2005) 'Social capital, networks, and knowledge transfer', *Academy of Management Review*, 30 (1), 146–65.

Isard, W. (1956) *Location and Space Economy*, Cambridge, MA: MIT Press.

Isard, W. (1960) *Methods of Regional Analysis: An Introduction to Regional Science*, Cambridge, MA: MIT Press.

ISTAT (1996) *Rapporto annuale: La situazione del Paese nel 1995*, Roma: Istituto Poligrafico e Zecca dello Statu.

IVE (Instituto Valenciano de Estadística) (2002) *Estadística Industrial de la Comunidad Valenciana 1999*, Valencia: Generalitat Valenciana.

Jarillo, J. C. (1988) 'On strategic networks', *Strategic Management Journal*, 9, 31–41.

Johanson, J. and Mattson, L. G. (1988) 'Internationalisation in industrial systems: a network approach', in N. Hood and J. E. Valhlen (eds), *Strategies in Global Competition*, London: Grom Helm.

Johanson, J. and Vahlne, J. (1990) 'The mechanism of internationalisation', *International Marketing Review*, 7 (4), 11–24.

Johanson, J. and Vahlne, J.-E. (2009) 'The Uppsala internationalization process model revisited: from liability of foreignness to liability of outsidership', *Journal of International Business Studies*, 40, 1411–31.

Johnson, G., Scholes, K. and Whittington, R. (2007) *Exploring Corporate Strategy*, Prentice Hall.

Jones, C., Hesterly, W. and Borgatti, S. (1997) 'A general theory of network governance: exchange conditions and social mechanisms', *Academy of Management Review*, 22 (4), 911–45.

Jones, R. M. (2002) *The Apparel Industry*, Oxford: Blackwell.

Jones, R. M. and Hayes, S. G. (2004) 'The UK clothing industry: extinction or evolution?', *Journal of Fashion Marketing and Management*, 8 (3), 262–78.

Joshi, M. P., Kathuria, R. and Porth, S. J. (2003) 'Alignment of strategic priorities and performance: an integration of operations and strategic management perspective', *Journal of Operations Management*, 21 (3), 353–69.

Julien, P.-A. (1996) 'Globalization: different types of small business behaviour', *Entrepreneurship & Regional Development*, 8, 57–74.

Kathuria, R. and Porth, S. J. (2003) 'Strategy-managerial characteristics alignment and performance: a manufacturing perspective', *International Journal of Operations & Production Management*, 23 (3), 255–76.

Kilduff, P. and Chi, T. (2006) 'Longitudinal patterns of comparative advantage in the textile complex', 10 (1–2), 134–68.

Knoke, D. and Kuklinski, J. H. (1983) *Network Analysis*, Beverley Hills, CA: Sage.

Kogut, B. (2000) 'The network as knowledge generative rules and the emergence of structure', *Strategic Management Journal*, 21, 405–25.

Kotha, P. and Orne, D. (1989) 'Generic manufacturing strategies: a conceptual synthesis', *Strategic Management Journal*, 10 (3), 211–31.

Kotha, S. and Swamidass, P. M. (2000) 'Strategy, advanced manufacturing technology and performance: empirical evidence from US manufacturing firms', *Journal of Operations Management*, 18, 255–77.

Krugman, P. (1991a) 'Increasing returns and economic geography', *Journal of Political Economy*, 99, 483–99.

Krugman, P. (1991b) 'History and industrial location: the case of the manufacturing belt', *American Economic Review, Papers and Proceedings*, 81, 80–3.

Krugman, P. (1993), 'First nature, second nature and metropolitan location', *Journal of Regional Science*, 33, 129–44.

Krugman, P. and Venables, A. (1995) 'Globalization and the inequality of nations', *Quarterly Journal of Economics*, 110 (4), 857–80.

Lainé, F. (2000) 'Agglomérations spécialisées d'établissements et systèmes localisés de production: une approche statistique', *Document de travail núm. H001*, Paris: INSEE.

Langlois, R. N. and Robertson, P. L. (1995) *Firms, Markets and Economic Change: A Dynamic Theory of Business Institutions*, New York: Routledge.

La Porta, R., Lopez-de-Silanes, F., Shleifer, A. and Vishny, R. (1997) 'Trust in large organizations,' *American Economic Review*, 87 (2), 333–8.

Lazerson, M. H. and Lorenzoni, G. (1999) 'The firms that feed industrial districts: a return to the Italian source', *Industrial and Corporate Change*, 8 (2), 235–55.

Lee, W. and Brash, J. (1978) 'The adoption of export as an innovative strategy', *Journal of International Business Studies*, 9 (1), 85–94.

Linares, E. and Pla, J. (2009) 'El offshoring en la empresa multinacional europea', *Revista Contabilidad y Tributación del Centro de Estudios Financieros*, 319, 125–76.

Lohrke, F., Bedeian, A. and Palmer, T. B. (2004) 'The role of top management teams in formulating and implementing turnaround strategies: a review and research agenda', *International Journal of Management Reviews*, 5–6 (2), 63–90.

Lösch, A. (1939) *The Economics of Location*, New Haven, CT: Yale University Press.

Lye, A. and Hamilton, R. T. (2001) 'Importer perspectives on international exchange relationships', *International Business Review*, 10, 109–28.

Maccarini, M. E., Scabini, P. and Zucchella, A. (2003a) 'Internationalisation strategies in Italian district-based firms: theorical modeling and empirical evidence', Paper presented at the Conference in Honour of Professor S. Brusco. Online, available: www.economia.unimo.it/convegni_seminari/CG_sept03/naviga.html (accessed 31 August 2009).

Maccarini, M. E., Scabini, P. and Zucchella, A. (2003b) 'International entrepreneurship and location-specific factors: a study on born global firms', paper presented at the 29th EIBA Conference, 11–13 December, Copenhagen.

McCormick, D. (1999) 'African enterprise clusters and industrialisation: theory and reality', *World Development*, 27 (9), 1531–51.

McEvily, B. and Zaheer, A. (1999) 'Bridging ties: a source of firm heterogeneity in competitive capabilities', *Strategic Management Journal*, 20 (2), 1133–58.

McGahan, A. M. and Porter, M. (1997) 'How much does industry matter, really? *Strategic Management Journal*, 18, 15–30.

Majocchi, A. (2000) 'Are industrial clusters going international? The case of Italian SMEs in Romania', Working Paper Univ. dell'Insubria 12. Online, available: http://eco.uninsubria.it/dipeco/Quaderni/files/QF2000_12.pdf (accessed 31 August 2009).

Marchi, F. (2005) 'The EU-25 textile & clothing industry in the year 2004', *General Assembly Euratex – June 2005*. Online, available: www.euratex.org (accessed 1 February 2010).

Mariotti, I. (2004) 'Internationalisation: threat or opportunity for the survival of the Italian district model?', IWSG Working Papers 9. Online, available: www.geo.uni-frankfurt.de/WSG/forschung/iwsg/iwsg_pdf/iwsg_working_paper_9_2004.pdf (accessed 31 August 2009).

Markusen, A. (1996) 'Sticky places in slippery space: a tipology of industrial districts', *Economic Geography*, 72, 293–313.

Marques, H. (2005) 'The skilled U-shaped Europe: is it really and on which side does it stand?', *Applied Economics*, 37 (19), 2205–20.

Marques, H. (2008) 'Trade and factor flows in a diverse EU: what lessons for the eastern enlargement(s)?', *Journal of Economic Surveys*, 22 (2), 364–408.

Marques, H. and Metcalf, H. (2009) 'Manufacturing wages in the enlarged EU: the role of neighbour-country effects', *Manchester School*, 77 (1), 65–81.

Marshall, A. (1890) *Principles of Economics*, London: Macmillan.

Marshall, A. (1919) *Industry and Trade*, London: Macmillan.

Martínez, M. T. (2001) *El modelo explicativo de la competitividad de la empresa en los distritos industriales: el efecto de los recursos compartidos*, unpublished thesis, Universitat Jaume I, Castellón.

Meardon, S. (2000) 'Eclecticism, inconsistency and innovation in the history of geographical economics', *History of Political Economy*, 32, 325–59.

Menguzzato, M. (1992) *La cooperación empresarial: análisis de su proceso*, Valencia: IMPIVA.

Miles, R. E. and Snow, C. C. (1992) 'Causes of failure in network organizations', *California Management Review*, 34 (4), 53–72.

Mittelstaedt, J. D. and Ward, W. A. (2003) 'Location, firm size and international trade: simultaneous measurement of the effects of internal and external scale economies on exporting', CUCIT Working Paper 0030115. Online, available: http://business.clemson.edu/cit/Documents/Locationfirmsize&internationaltrade%20working%20paper.pdf (accessed 31 August 2009).

Molina, F. X. (2001a) 'European industrial districts: influence of geographic concentration on performance of the firm', *Journal of International Management*, 7, 1–18.

Molina, F. X. (2001b) 'Human capital in the industrial districts', *Human System Management*, 20 (4), 319–32.

Molina, F. X. (2002) 'Industrial districts and innovation: the case of the Spanish ceramic tiles industry', *Entrepreneurship & Regional Development* (14), 317–35.

Molina, F. X. (2005) 'The territorial agglomerations of firms: a social capital perspective from the Spanish tile industry', *Growth and Change*, 36 (1), 74–99.

Molina, F. X. and Bou, J. C. (2000) 'Un análisis del desempeño empresarial en el distrito industrial a partir de la variedad estratégica', *Revista Europea de Dirección y Economía de la Empresa*, 9 (2), 9–20.

Molina, F. X. and Martínez, M. T. (2003) 'The impact of industrial district affiliation on firm value creation', *European Planning Studies*, 11 (3), 155–70.

Molina, F. X. and Martínez, M. T. (2004) 'How much difference is there between industrial district firms? A net value creation approach', *Research Studies*, 33, 473–86.

Möller, K. and Svahn, S. (2003) 'Managing strategic nets: a capability perspective', *Marketing Theory*, 3 (2), 201–26.

Myro, R. and Fernández-Otheo, C. (2004) 'La deslocalización de empresas en España: La atracción de la Europa central y oriental', *Información Comercial Española*, 818, 185–204.

Myro, R. and Gandoy, R. (2001) 'Sector industrial', in J. L. García Delgado (ed.), *Lecciones de Economía Española*, Madrid: Civitas.

Nassimbeni, G. (2003) 'Local manufacturing system and global economy: are they compatible? The case of the Italian eyewear district', *Journal of Operations Management*, 21, 151–71.

North, D. (1991) 'Institutions', *Journal of Economic Perspectives*, 5 (1), 97–112.

Nunnally, J. C. and Bernstein, I. H. (1994) *Psycometric Theory*, New York: McGraw Hill.

O'Donoghue, D. and Gleave, B. (2004) 'A note on methods for measuring industrial agglomeration', *Regional Studies*, 34, 419–27.

OECD (1992) *Structural Change and Industrial Performance: A Seven Country Growth Decomposition Study*, Paris: OECD.

Oliver, A. L. (1998) 'Networking network studies: an analysis of conceptual configurations in the study of inter-organizational relationships', *Organization Studies*, Fall, 1–36.

Omiccioli, M. and Quintiliani, F. (2000) 'Assetti imprenditoriali, organizzazione del lavoro e mobilità nei distretti industriali' in L. F. Signorini (ed.), *Lo Sviluppo Locale: Un'indagine della Banca d'Italia sui Distretti Industriali*, Roma: Meridiana.

Ostrom, E. (1996) 'Crossing the great divide: coproduction, synergy, and development,' *World Development*, 24 (6), 1073–87.

Overby, J. W. and Servais, P. (2005) 'Small and medium-sized firms, import behavior: the case of Danish industrial purchasers', *Industrial Marketing Management*, 34, 71–83.

Oxelheim, L. and Ghauri, P. (2004) *European Union and the Race for Foreign Direct Investment in Europe*, Oxford: Elsevier.

Paniccia, I. (1998) 'One, a hundred, thousands of industrial districts: organizational variety of local networks of small and medium-sized enterprises', *Organization Studies*, 4 (19), 667–99.

Paniccia, I. (1999) 'The performance of industrial districts: some insights from the Italian case', *Human Systems Management*, 18 (2), 141–59.

Paniccia, I. (2002) *Industrial Districts: Evolution and Competitiveness in Italian Firms*, Cheltenham: Edward Elgar.

Parr, J. B. (2002a) 'Agglomeration economies: ambiguities and confusions', *Environment and Planning*, 34, 717–31.

Parr, J. B. (2002b) 'Missing elements in the analysis of agglomeration economies', *International Regional Science Review*, 25 (2), 151–68.

Pavitt, K. (1984) 'Sectoral patterns of technical change: towards a taxonomy and a theory', *Research Policy*, 6 (13), 343–73.

Perrow, C. (1992) 'Small-firm networks', in N. Nohria and R. G. Eccles (eds), *Networks and Organizations: Structure, Form and Action*, Boston, MA: Harvard Business School Press.

Pietrobelli, C. (2000) 'Competitiveness and its socio-economic foundations: empirical evidence on the Italian industrial districts', in M. Bagella and L. Becchetti (eds), *The Competitive Advantage of Industrial Districts*, New York: Physica-Verlag.

Pietrobelli, C. and Olarte, T. (2002) 'Enterprise clusters and industrial districts in Colombia's fashion sector', *European Planning Studies*, 10 (5), 541–62.

Piore, M. J. (1992) 'Obra, trabajo y acción: experiencia de trabajo en un sistema de producción flexible', in F. Pyke, G. Becattini and W. Sengenberger (eds), *Los Distritos Industriales y las Pequeñas Empresas*, Madrid: Ministerio de Trabajo and Seguridad Social.

Piore, M. J. and Sabel, C. F. (1984) *The Second Industrial Divide*, New York: Basic Books.

Piqueras, J. (1999) *El espacio valenciano: Una síntesis geográfica*, Valencia: Editorial Gules.

Pla, J. and León, F. (2004) *Dirección de Empresas Internacionales*, Madrid: Prentice-Hall.

Pla, J. and Puig, F. (2009) 'Is the influence of the industrial district on international activities being eroded by globalization? Evidence from a traditional manufacturing industry', *International Business Review*, 18 (5), 435–45.

Pla, J., Puig, F. and Linares, E. (2007) 'Crisis, actitudes directivas and estrategia en los sectores manufactureros tradicionales: el sector textil español', *Universia Business Review*, 14, 68–83.

Porter, M. (1990) *The Competitive Advantages of the Nations*, New York: The Free Press.

Porter, M. (1998) 'Clusters and the New Economic Competition', *Harvard Business Review*, 76 (6), 77–90.

Puig, F., Marques, H. and Ghauri, P. (2009) 'Globalization and its impact on operational decisions: the role of industrial districts in the textile industry', *International Journal of Operations and Production Management*, 29 (7), 692–719.

Pyke, F. and Sengenberger, W. (1992), *Industrial Districts and Local Economic Regeneration*, Geneva: International Institute for Labour Studies.

Pyke, F., Becattini, G. and Sengenberger, W. (1990) *Industrial Districts and Inter-Firm Co-operation in Italy*, Geneva: International Labour Organisation.

Rabellotti, R. (1995) 'Is there an 'industrial district model'? Footwear districts in Italy and Mexico compared', *World Development*, 23 (1), 29–41.

Rabellotti, R. (2003) 'How globalisation affects Italian industrial districts: the case of Brenta', Paper presented at the RSA International Conference. Online, available: www.regional-studies-assoc.ac.uk/events/pisa03/rabellotti.pdf (accessed 31 August 2009).

Rauch, J. (1999) 'Networks versus markets in international trade', *Journal of International Economics*, 48 (1), 7–35.

Rauch, J. and Trindade, V. (2003) 'Information, international substitutability and globalization', *American Economic Review*, 93, 775–91.

Rialp, A. and Rialp, J. (2001) 'Conceptual framework on SME's internationalisation: past, present and future trend of research', in C. N. Axinn and P. Mathyssens (eds), *Reassessing the Internationalisation of the Firm*, Oxford: Elsevier.

Robbins, D. and Pearce, J. (1992) 'Turnaround: retrenchment and recovery', *Strategic Management Journal*, 13 (4), 287–309.

Roberts, M. and Tybout, J. (1997), 'An empirical model of sunk costs and the decision to export,' *American Economic Review*, 87 (4), 545–64.

Rodrik, D. (1998), 'Symposium on globalization in perspective: an introduction', *Journal of Economic Perspectives*, 12 (4), 3–8.

Sachwald, F. (1998) 'Cooperative agreements and the theory of the firm: focusing on barriers to change,' *Journal of Economic Behavior & Organization*, 35 (2), 203–25.

Sammarra, A. and Belussi, F. (2006) 'Evolution and relocation in fashion-led Italian districts: evidence from two case-studies', *Entrepreneurship and Regional Development*, 18, 543–62.

Santa María, M. J. (1999) *Pequeñas empresas y distritos industriales: el textil confección en la Comunidad Valenciana*, Institut de Cultura Juan Gil-Albert, Colección Textos Universitaris, Generalitat Valenciana.

Santa María, M. J., Giner, J. M. and Fuster, A. (2004) 'The identification of the local systems in Spain: a new approach', *European Congress of the Regional Science Association*. Online, available: www.ersa.org/ersaconfs/ersa04/PDF/122.pdf (accessed 31 August 2009).

Saxenian, A. (1994) *Regional Advantage: Culture and Competition in Silicon Valley and Route 128*, Cambridge, MA: Harvard University Press.

Schmitz, H. (1995) 'Collective efficiency: growth path for small-scale industry', *Journal of Development Studies*, 31 (4), 529–66.

Schmitz, H. (2000) '¿Tiene importancia la cooperación local? Experiencias de clusters industriales en el Sur de Asia and América Latina', *El Mercado de Valores*, September, 4–17.

Sforzi, F. (1990) 'Marshallian industrial districts in the Italian economy', in F. Pyke, G. Becattini and W. Sengenberger (eds), *Industrial Districts and Local Economic Regeneration*, Geneva: International Institute for Labor Studies.

Sforzi, F. (2002) 'The industrial district and the 'new' Italian economic geography', *European Planning Studies*, 10 (4), 439–47.

Signorini, L. F. (1994a) 'Una verifica quantitativa dell'effetto distretto', *Svilupo Locale*, 1 (1), 31–70.

Signorini, L. F. (1994b) 'The price of Prato, or measuring the industrial district effect', *Papers in Regional Science*, 73 (4), 369–92.

Skinner, W. (1969) 'Manufacturing missing link in corporate strategy', *Harvard Business Review*, 47, 136–45.

Smith, A., Pickles, J., Begg, R., Roukova, P. and Bucek, M. (2005) 'The future of the textiles and clothing sector in the enlarged textiles and clothing industry: reflections on the European Commission's outward processing, EU enlargement and regional relocation in the enlarged European Union', *European Urban and Regional Studies*, 12, 83–91.

Snodgrass, C. R. (1993) 'The use of networks in cross border competition', *Long Range Planning*, 26 (2), 41–50.

Soler, V. (2000) 'Verificación de las hipótesis del distrito industrial: Una aplicación al caso valenciano', *Economía Industrial*, 334, 13–23.

Soler, V. and Hernández, F. (2001) 'La misurazione delle economie esterne marshalliane atraverso i modelli DEA', *Sviluppo Locale*, 8 (16), 86–105.

Sölvell, Ö. (2009) *Clusters: Balancing Evolutionary and Constructive Forces*, Stockholm: Ivory Tower Publishers.

Spies, J. and Marques, H. (2009), 'Trade effects of the Europe Agreements: a theory-based gravity approach', *Journal of International Trade and Economic Development*, 18 (1), 11–35.

Staber, U. (1997) 'Specialization in a declining industrial district', *Growth and Change*, 28, 475–95.

Staber, U. (1998) 'Inter-firm co-operation and competition in industrial districts', *Organization Studies*, 19 (4), 701–24.

Staber, U. (2001) 'Spatial proximity and firm survival in a declining industrial district: the case of knitwear firms in Baden-Württemberg', *Regional Studies*, 35 (4), 329–41.

Staber, U. (2007) 'Contextualizing research on social capital in regional clusters', *International Journal of Urban and Regional Research*, 31 (3), 505–21.

Staber, U. and Morrison C. (1999) 'The empirical foundations of industrial district theory', paper presented at the ISRN Workshop on Globalization and Regional Innovation Systems, 17–19 May, Toronto.

Steinle, C. and Schiele, H. (2002) 'When do industries cluster? A proposal on how to assess an industry's propensity to concentrate at a single region or nation', *Research Policy*, 31, 849–58.
Stengg, W. (2001) 'The textile and clothing industry in the EU: a survey', Enterprise Papers, 2001/02, Enterprise Directorate-General of the European Commission.
Stevens, J. (1996) *Applied Multivariate Statistics for the Social Sciences*, Mahwah, NJ: Lawrence Erlbaum.
Storper, M. (1997) *The Regional World*, New York: Guilford Press.
Such, J. (1995) 'Estrategias comerciales en los distritos industriales: El caso de la industria textil de Alcoi-Ontinyent', *Investigación Comercial Española*, 746, 57–65.
Sura, W. (2004) 'The textile and clothing in the UE', *Statistics in Focus*. Online, available: http://epp.eurostat.cec.eu.int/portal/page (accessed 1 February 2010).
Szarka, J. (1990) 'Networking and small firms', *International Small Business Journal*, 8 (2), 10–22.
Taplin, I. (2006) 'Restructuring and reconfiguration: the EU textiles and clothing industry adapts to change', *European Business Review*, 18 (3), 172–86.
Tattara, G. (2001) 'L'efficienza dei distretti industriali: una ricerca condotta dal servizio studi della Banca d'Italia', *Economia e Società Regionale*, 4, 114–43.
Teece, D. J., Rumelt, R., Dosi, G. and Winter, S. (1994) 'Understanding corporate coherence: theory and evidence', *Journal of Economics Behavior and Organization*, 23, 1–30.
Thorelli, H. B. (1986) 'Networks: between markets and hierarchies', *Strategic Management Journal*, 7, 37–51.
Tomás, J. A. and Torrejón, M. (2000) 'Innovación y sistemas territoriales de PME en sectores tradicionales: la industria textil valenciana', *Economía Industrial*, 331, 151–66.
Toni, A. D. and Nassimbeni, G. (2003) 'Small and medium district enterprises and the new product development challenges: evidence from Italian eyewear district', *International Journal of Operations and Production Management*, 23 (6), 678–97.
Toyne, B., Arpan, J. S., Barnett, A. H., Ricks, D. A. and Shimp, T. A. (1984) *The Global Textile Industry*, London: George Allen & Unwin.
Trullén, J. (1990) 'Caracterización de los distritos industrials: el distrito industrial marshalliano en el debate actual sobre desarrollo regional y localización industrial', *Economía Industrial*, 273, 151–63.
UNCTAD (2004) 'Service offshoring takes off in Europe – in search of improved competitiveness', Paris: UNCTAD.
Vázquez, A (1999) *Desarrollo, redes e innovación: Lecciones sobre desarrollo endógeno*, Madrid: Pirámide.
Venables, A. (1999) 'The international division of industries: clustering and comparative advantage in a multi-industry model', *Scandinavian Journal of Economics*, 101, 495–513.
Vernon, R. (1966) 'International investment and international trade in the product cycle', *Quarterly Journal of Economics*, 80 (1), 190–207.
Visser, E. J. (1999) 'A comparison of clustered and dispersed firms in the small-scale clothing industry of Lima', *World Development*, 27 (9), 1553–70.
von Thünen, J. (1826) *The Isolated State*, n.p: n.p.
Weber, A. (1910) *Theory of Location of Industries*, Chicago, IL: University of Chicago Press.
Weber, A. (1929) *Theory of the Location of Industries*, Chicago, IL: University of Chicago Press.

Welch, D. E. and Welch, L. S. (1996) 'The internationalization process and networks: a strategic management perspective', *Journal of International Marketing*, 4 (3), 11–28.

Welch, L. S. and Luostarien, R. (1988) 'Internationalization: evolution of a concept', *Journal of General Management*, 14, 34–55.

Wiklund, J. and Karlsson, C. (1994) 'Flexible companies in an industrial district: the case of the Gnosjö region in Sweden', in J. M. Veciana (ed.), *SMEs: Internationalization, Networks and Strategy*, Aldershot: Avebury.

Williamson, J. G. (1996) 'Globalization, convergence, and history', *Journal of Economic History*, 56 (2), 277–306.

Williamson, J. G. (1997) 'Globalization and inequality, past and present', *World Bank Research Observer*, 12 (2), 117–35.

Williamson, J. G. (1998) 'Globalization, labour markets and policy backlash in the past', *Journal of Economic Perspectives*, 12 (4), 51–72.

Williamson, O. E. (1979) 'Transaction cost economics: the governance of contractual relations', *Journal of Law and Economics*, 22, 223–61.

Woolcock, M. and Narayan, D. (2000) 'Social capital: implications for development theory, research, and policy,' *World Bank Research Observer*, 15 (2), 225–49.

Wooldridge, J. M. (2002) 'Econometric analysis of cross section and panel data', Cambridge, MA: MIT Press.

Ybarra, J. A. (1991) 'Determinación cuantitativa de distritos industriales: la experiencia del País Valenciano', *Estudios Territoriales*, 37, 55–67.

You, J.-L. and Wilkinson, F. (1994) 'Competition and co-operation toward understanding industrial districts', *Review of Political Economy*, 6 (3), 259–78.

Zahra, S. A. and George, G. (2002) 'International entrepreneurship: the current status of the field and future research agenda' in M. Hitt, R. Ireland, M. Camps and D. Sexton (eds), *Strategic Leadership: Creating a New Mindset*, London: Blackwell.

Zhao, H. and Zou, S. (2002) 'The impact of industry concentration and firm location on export propensity and intensity: an empirical analysis of Chinese manufacturing firms', *Journal of International Marketing*, 10 (1), 52–71.

Zourek, H. (2007) 'The European Commission's new industrial policy in an integrating and globalizing world', *Journal of Industry, Competition and Trade*, 7, 285–95.

Zucchella, A. (2002) 'Born global versus gradually internationalizing firms: an analysis based on the Italian case', paper presented at the 28th EIBA Conference, Athens, 8–10 December.

Zucchella, A. (2006) 'Local cluster dynamics: trajectories of mature industrial districts between decline and multiple embeddedness', *Journal of Institutional Economics*, 2 (1), 21–44.

Index

Note: Page numbers in *italics* denote tables, those in **bold** denote figures.